MYSTERIES AND SECRETS OF TIME

LIONEL AND PATRICIA FANTHORPE

MYSTERIES AND
SECRETS OF TIME

THE DUNDURN GROUP
TORONTO

Copy-editor: Andrea Waters
Typesetting: Alison Carr
Printer: Marquis

Library and Archives Canada Cataloguing in Publication

Fanthorpe, R. Lionel
    Mysteries and secrets of time / Lionel and Patricia Fanthorpe.

ISBN 978-1-55002-677-1

    1. Time--Popular works.   2. Space and time--Popular works.   3. Time travel--Popular works.   I. Fanthorpe, Patricia   II. Title.

BD638.F35 2007   115   C2007-900908-5

1   2   3   4   5   11   10   09   08   07

 Conseil des Arts   Canada Council
du Canada   for the Arts   Canadä

ONTARIO ARTS COUNCIL
CONSEIL DES ARTS DE L'ONTARIO

We acknowledge the support of the Canada Council for the Arts and the Ontario Arts Council for our publishing program. We also acknowledge the financial support of the Government of Canada through the Book Publishing Industry Development Program and The Association for the Export of Canadian Books, and the Government of Ontario through the Ontario Book Publishers Tax Credit program, and the Ontario Media Development Corporation.

Care has been taken to trace the ownership of copyright material used in this book. The author and the publisher welcome any information enabling them to rectify any references or credits in subsequent editions.

                                                                                    J. Kirk Howard, President

www.dundurn.com

Dundurn Press   Gazelle Book Services Limited   Dundurn Press
3 Church Street, Suite 500   White Cross Mills   2250 Military Road
Toronto, Ontario, Canada   High Town, Lancaster, England   Tonawanda, NY
M5E 1M2   LA1 4XS   U.S.A. 14150

This book is dedicated to all our family and friends, and to our TV viewers, radio listeners, lecture audiences, and readers, who are so kind, helpful, and supportive about our research into these strange unsolved mysteries — and who always encourage us to go out and investigate more!

Patricia & Lionel Fanthorpe
www.Lionel-Fanthorpe.com

TABLE OF CONTENTS

FOREWORD

ONCE AGAIN, Lionel and Patricia Fanthorpe have asked me to write a foreword for their new book. Some forty books over the last years have come from their teeming brains, and each and every one of the forewords has been entrusted to me. We have grown old together!

This, their latest investigative book, probes into the secrets and vagaries of time. Early on in the book, they evoked a cherished memory for me. Though now, sadly, I am a very old man of ninety-three, this memory is from my early years as an undergraduate at Oxford. It was sometime in the early 1930s, when I saw notice that the renowned scholar Professor Einstein was to give a public lecture on his theory of relativity. I joined the large gathering anxious to see and hear the great man. His appearance alone would have been worth the effort with his large head of flamboyant white hair and his penetrating eyes. On the other hand, almost all that he said was totally beyond me. The stark truth was that my capacity to reason was light-years away from his. In company with many others in the room, no doubt, we might have been comfortable, to some degree at least, with a talk on Shakespeare, or Goethe, or Victor Hugo, or any such other writer or thinker, but faced with talk of quantum theory and wave particle dualities we were hopelessly lost. I returned to college that evening a sadder but not a wiser man knowing I had been in the presence of a truly great man with an intellect and a realm of scholarship I, for one, could never share. The Albert Einsteins, the Isaac Newtons, the Stephen Hawkings are all exceptional people and light-years ahead of even the best of us.

At first reading, certainly in the early chapters, the Fanthorpes seem to do an Einstein on us! You may be tempted to surrender before their intellectual demands. The metaphysics of time may not be for us, neither its philosophy nor its theology. We may rest content to leave such depths to the intellectuals and simply accept time as it comes, without question, be it short or long, fraught or untroubled, and generally make the best of what we have. The adage "Eat, drink, and be merry, for tomorrow you die" may not be an admirable philosophy, but at least it's one we can understand.

But don't be defeated. Read on. Time as a concept may be beyond us, but inside time itself there are lesser mysteries that are within our experience and understanding. Here, too, the Fanthorpes do their best for us, sharing with us, as they see them, some of time's smaller and intriguing problems.

We are aware of an occasional experience that has attracted the term *déjà vu*. Time seems to repeat itself. We have known what it is to visit an area where, to our certain knowledge, we have never been before. Yet it is strangely and immediately familiar to us. It can equally be true of a house we are invited to for the first time. We know, somehow, what lies around the corner of the village, or where all the rooms are. We have to wonder if we were there long years ago when too young to be consciously aware of our surroundings. Or did we know of it from a previous phase of life? Lionel tests his own life to see if such earlier existences are possible. Under hypnosis he seems to recover five earlier lives and could, perhaps, have recovered others. He doesn't know whether his regressions are real or fancied. Did he, under hypnosis, recall a character and lifestyle he would have enjoyed, given the chance? Knowing Lionel as I do, I would have expected him to be a musketeer at Waterloo, or a Templar knight at the Crusades, but not in a million years to see himself as a Franciscan monk, devoted to the solitary life of a monastery. Readers will take a view for themselves, accepting or rejecting any thought of reincarnation.

The authors describe cases of individuals seeming to override the natural boundaries of time. They appear as if from nowhere. One such the Fanthorpes describe for us is Kaspar Hauser, appearing in Nuremberg in 1828 with no apparent background and almost no language, dying as strangely as he was born. Who needs fictional Dr. Who with his time machine, if individuals can themselves cross in and out of the normal barriers of time?

Time seems also, if we go where the Fanthorpes lead, to have lapses or glitches. Certainly, they give many intriguing illustrations of it. One such they gave in one of their earlier books. It concerns a destroyer in the last war, the USS *Eldridge*. In 1942, it was at anchor in Delaware Bay, subject to experiments, the aim of which was to increase the invisibility of the ship in the face of an enemy. It seems to have worked even better than was planned as the destroyer then turned up in far away Norfolk, Virginia, to the pronounced physical discomfort of the crew — jet lag on a massive scale!

Most of us, who live in time, don't much care if it's linear or infinite or cyclic or anything else. To us it seems rhythmic, unstoppable, remorseless even, rolling along, bearing us with it, with all the indifference of the Mississippi River of the old song from *Show Boat*. But if there are anomalies, lapses, and slips to be found, at least we ought to know about them, and we are grateful to the authors for their diligent and painstaking research. Conclusions drawn will be up to us.

At my age now time must be running to its end. The body weakens; the intellect finds it harder to keep its concentration. I doubt if I shall write any more forewords, but I know the Fanthorpes will write many more books. I wish them well as they continue to understand and share with us some of the bizarre twists and turns of life. They have unearthed many strange mysteries, introduced us to some weird and wonderful people, proposing explanations often difficult for us, with less fertile imaginations, to accept. Their books deserve and have a wide circulation. I hope this one will join the best of them.

Canon Stanley Mogford, M.A.
Cardiff 2006

(The authors, as always, are deeply grateful to Canon Mogford, rightly acknowledged to be one of the greatest scholars in Wales, for his tremendous help, encouragement, and friendship.)

INTRODUCTION

TIME IS ONE of the strangest and most mysterious concepts with which the human mind has to contend. Physicists and philosophers, thinkers and theologians, magi, metaphysicians, and mathematicians have all made their distinctive attempts to explain at least part of the enigma of time and its relationship with 3-D space. Some of these intellectual explorers have barely scratched the riddle's elusive surface; others have penetrated more deeply, but only in their own specialized and limited areas.

Time raises many questions. Is it linear? Is it circular? Is it infinite, or does it have a definite beginning and end? Are its nature and function unchangeable and invulnerable, or can vast inputs of matter and energy affect them? How do black holes and white vortices link up with the conundrum of time? Is it fixed while everything else drifts past, or does time itself do the flowing? Is our human experience of time — or what we *think of* as our experience of time — a genuine, external reality? Is time objective or subjective?

Our long years of research and first-hand investigations into the anomalous, the paranormal, and the unsolved have frequently involved reports of apparent time travel; strange warps and twists in time; déjà vu; reincarnation; and near-death experiences. We have also analyzed and examined curious cases of apparent longevity: the biblical patriarchs like Methuselah, the enigmatic Count of St. Germain, and various inscrutable, ancient yoga masters who appear to have lived for centuries by acquiring the power to slide in and out of time at will.

Other aspects of the mystery of time concern strangely perceptive people — we might even call them prophets — like the Brahan Seer, Mother Shipton, and Nostradamus. Allegedly, they looked more penetratingly than most of us through the hazy mists of time and wrote down — sometimes ambiguously — the disturbing enigmas they thought they could see there.

There is also the time-oriented mystery of people like Leonardo da Vinci, so gifted that they seem to have been born centuries *ahead* of their time. Is it possible that they had mastered time travel and were getting their ideas from some future probability track? Or had some strange glitch in time brought a mind from the future into a Renaissance body?

We have also examined inexplicable incidents such as those reported from York and from Wroxham Broad in Norfolk, where Roman legions were said to have been seen marching in procession many centuries after the Romans left Britain. We have looked closely into the case of the two English schoolteachers who visited Versailles at the dawn of the twentieth century and saw things in those picturesque French gardens that made them believe they were back in the eighteenth.

Scientists have put forward numerous ideas about the possibility of actual, physical time travel using machines of various types like those that science fiction writers have used ever since H.G. Wells wrote *The Time Machine*. Other investigators have considered the possibility that time travel could involve out-of-body experiences so that time travellers' discarnate minds might find themselves occupying different bodies in another time and space. Some psychical researchers wonder whether the frequently reported glowing orb phenomenon could be connected with this theory.

It has also been speculated that time forms strange, inter-dimensional bridges or portals that open into probability tracks — the mystifying Worlds of If.

Other clues to time's anomalous behaviour are buried among the anachronistic artifacts that turn up from time to time: something that looked for all the world like a modern automobile's spark plug was found among ancient fossils; a modern-looking zinc and silver alloy vase was dug from a 100,000-year-old rock stratum near Boston, Massachusetts; a gold chain was found inside a piece of coal. The Piri Re'is map has never been definitively explained, and Dr. Cabrera's discoveries among ancient stones in Peru are not easy to explain either — why do those ancient carvings show modern instruments like telescopes?

It has seemed to us in the course of our extensive research into these strangely associated teleological mysteries that there is a unified field theory of time — and we present that theory as our conclusion. Bringing the various riddles together in this way creates a chain reaction and makes one enigma serve as a vital clue to the next.

Chapter 1

THE SCIENCE OF TIME

SCIENCE IS not quite a religion; not quite a philosophy; not quite a lifestyle. To approach the science of time, it may be useful first to attempt to analyze and define science itself. It can be thought of as an attitude towards our environment; a way of examining our surroundings; a technique for trying to find out *who* we are, *where* we are, and *why* we are. A time-scientist would also suggest that science is concerned with *when* we are, although time-scientists do not regard that question in quite the same way that historians do. Science is interested in causes and effects and the relationships between them. The true scientist is the woman or man who observes objects and events as carefully as possible, and then creates theories and hypotheses that attempt to understand and explain those phenomena.

Some of the strangest phenomena associated with time are the accounts of apparent time slips and déjà vu experiences, which are dealt with in detail in later chapters. Whatever time really is, it appears to be subject to warping — perhaps even to rupturing and rejoining — and honest, objective accounts of these events must not be ignored.

Theoretical and empirical scientists look for the links that hold objects and events together in the right order: they realize that without an awareness of time it is impossible to establish such a sequential order. Science is one useful and practical method — the best so far devised — of studying cause and effect, but the scientist cannot measure and evaluate cause and effect unless and until he can measure and evaluate *time*. The paradox of time-science is that when we try to examine time, we are trying to examine something that is itself an essential part of the analytical tool kit. It is the classical dilemma of the blacksmith trying to make a hammer and anvil — while needing to use both to carry out the act of their creation.

A writer has to form a mental concept first, and then encode that idea via an appropriate language. Next, he must polish and refine the chosen imagery, vocabulary, grammar, and syntax into an optimum verbal expression. Finally, this reaches the reader — hopefully still carrying most of the writer's intended message within it. When this sequence is analyzed in terms of time-science, it becomes clear that the basic mental concept is formed at an earlier time than that at which the linguistic expression is finalized. It also becomes clear that in this sequence thought precedes encoding, which in turn precedes the reader's reception of the message and his subsequent decoding and understanding of it. Without this teleological sequencing, analysis of even the simplest and most direct thought projection is impossible. It is time that enables us to understand the order of events, and unless we can understand that order, we cannot begin to undertake the scientific quest for cause and effect. Time is, therefore, not simply one of many phenomena being studied — it is an essential component of the studying apparatus itself.

We begin, then, by considering the service that time renders to science, but although that sheds a little useful light on a small part of the problem, it reveals almost nothing about the actual nature of time. It tells us about what time *does*, but nothing about what time *is*. By examining time as the measuring rod for sequences of events, for causes and effects, we are confronted by the question of whether time can exist in an environment where nothing changes. If time is used to measure causes and effects, and to reveal in what order events occur, can time still exist in an environment where nothing is occurring?

Time-scientists are also concerned with the mystery of why time seems to be unidirectional. They talk of the time arrow, and of time's irreversibility. If time is merely the fourth dimension of a continuum in which the other three dimensions can be traversed easily in several directions, what gives time the properties of a restrictive, one-way diode rather than a highly conductive, multidirectional thick copper wire? Allowing that our human experiences and perceptions of time as unidirectional may bear some resemblance to the objective reality of time (if time actually *has* an objective reality), the time-scientist is confronted by the question of the relative degrees of "reality" belonging to the states of time other than the infinitesimally brief *now*: what we refer to as the past and the future.

If *now* has an objective reality, is that objective reality qualitatively similar to what has happened already and to what has not yet happened?

It cannot be quantitatively similar because of the incredible brevity of the moment. Is this amazing brevity a measurable absolute like the speed of light? Is there such a thing as an ultimately short unit of time? This is something we would like to call an instanton (although the term *instanton* has already been used in a different sense related to quantum mechanics and mathematical physics, where it is also referred to as a pseudoparticle; it is helpful in providing classical solutions to equations of motion with a finite, non-zero action). In this book we will use the word in our sense of its being the teleological equivalent of the smallest particle of matter. The alternative to instanton theory is that time is infinitely divisible: if that's true, no ultimately small instanton can exist because the shortest interval of time known and measured can be halved and halved again indefinitely as our technology improves.

The waves of the electromagnetic spectrum (including light) travel at a speed of approximately 300,000 kilometres (186,000 miles) per second. In our present state of knowledge that seems to be the universe's speed ceiling. By considering the ways in which that very high speed can be measured, we may arrive at theories of ways in which an instanton of time can be measured — if it exists.

During the early seventeenth century, the few scientists who were around didn't accept the idea that light had a speed. Their practical, common-sense experience of life told them that light didn't travel anywhere at any speed: it was just *there*. It was instantaneous. It could cover any distance in zero time. Galileo (1564–1642) — often called the father of physics, the father of astronomy, and the father of science — didn't feel happy

Galileo.

about the instantaneous light hypothesis and undertook an experiment to measure the speed of light. He stood on one hilltop while his assistant went to another summit a mile away. Both men carried lanterns with shutters. The idea was that Galileo would open his lantern shutter, and as soon as his assistant saw Galileo's light, he would open his shutter. Galileo intended to repeat the experiment once or twice to confirm their findings and their expected measurement of the time required for the light to travel a mile. There was nothing wrong with the broad outline of his methodology — Galileo failed simply because light was far, far too fast to be measured using any instruments available in his time. (It was a prime era for pioneering scientists: Galileo's great contemporary Johannes Kepler, the first modern astrophysicist, lived from 1571 to 1630.) An approximate time for light to cover one mile would be 0.000005 of a second. In order to get any sort of figure from such an experiment, it would be necessary to use a much greater distance than the one Galileo and his assistant used.

The man who did that was Ole Roemer (1644–1710), a Danish astronomer who was studying Io, one of Jupiter's many moons, in 1675. He noticed that Io's orbit could vary by as much as twenty minutes — which was a significant observation considering that Io completed an orbit in only one day and eighteen and a half hours. Roemer worked out that these twenty-minute differences he had observed were due to the speed of light. When the observer on Earth was farther away from Jupiter, the apparent increase in the orbit time was simply because the light took longer to travel from Io to Earth. When Jupiter and Earth were closer, the time difference went the other way. Roemer's calculations gave him a figure of 225,000 kilometres per second (against the modern 299,792, which is conveniently rounded up to 300,000 kilometres per second).

Time-scientists are understandably interested in the speed of light because of Einstein's work on time dilation, which will be examined in detail in Chapter 2. This is a curious phenomenon that indicates to a stationary observer that the rate at which time passes in objects that are moving relative to his stationary situation is slower. His stationary clock, for example, is recording two seconds, while an identical moving clock is recording only one second. In Einstein's theory of special relativity, clocks that are moving relative to an inertial system (the motionless observer) run more slowly. In Einstein's theory of general relativity, it is gravity, not movement, that makes clocks run more slowly. Clocks that are close to a massive body that has a strong gravitational field will run slower than clocks that are not influenced by the same gravitational strength.

The intriguingly named gravitational redshift is the phenomenon of light apparently losing energy as it moves away from a massive body, so that spectral lines shift towards the red end of the spectrum. The gravitational blueshift reverses the process: light coming from a zone of weaker gravity undergoes a shift of spectral lines towards the blue end of the spectrum.

The speed of light is, therefore, of primary importance to the scientist examining the phenomenon of time and the effect that movement has on it. If lightspeed *is* the absolute velocity limit in the known universe, what happens to time when lightspeed is reached?

Time dilation does not seem to be merely a subjective experience within the human mind. Time dilation has to be thought of as something real and objective if progress is to be made: it is not an illusion, and neither is it a malfunction of the observer's mind. To make any sense of the scientific mysteries and secrets of time, we have to begin with the fact that time dilation actually happens.

When we begin to consider the challenging questions that are associated with the enigma of time, it is necessary to decide which of them belong here in an assessment of the science of time, and which of them belong in an analysis of the philosophy and theology of time. An assessment of the reality of the past and future may not be a question that science can answer.

How is the apparent flow of time to be analyzed? Is it a relative thing? Imagine that in a cosmos containing nothing else, in which there are no reference points whatsoever, two perfectly spherical spaceships pass each other. They may be moving in opposite directions. Or one of them could be stationary, while the other is mobile. Or they may be travelling in the same direction at different velocities. This image provides a model for the "flow of time" concept.

If the human observer is envisaged as one of the spherical space-ships, and time is seen as the other, then the same possible explanations still apply. The observer could be stationary while time flows past. Time could be stationary while the observer and his experiential concatenation of observed events flow past. Or both time and the observer could be progressing at different velocities along the same track. The nature, velocity, and direction of time-flow ought to be susceptible to scientific analysis in due course — even if it can't be done definitively at our present level of scientific knowledge and technological skill.

Another challenging question concerns the magnitude of the past and future. In one sense, both may be thought of as infinite: yet, if neither is truly *real*, can they possess anything akin to magnitude? The non-existent must by its very nature be impossible to measure. A further consideration is that although the future may be both boundless and infinite, the past may have begun with the Big Bang. If there was a Big Bang, did time come into existence then — along with matter and 3-D space? Is there a beginning to time, and will it have an end? Or is time infinite?

In considering the challenging question of the objectivity or subjectivity of time, there is a vast quantity of exploration still to be undertaken in the field of neurology. What exactly is it in the human brain that experiences time? With 10^{14} neurons playing their sophisticated symphony of thought by making electrochemical connections inside the brain, it is not easy to discover the time-discerning process or to pin down its precise cranial location. This neurological treasure hunt is the kind of problem that science is best fitted to solve — and most assuredly will solve in due course.

As we shall show in depth in Chapter 5, some of the finest philosophical minds have lined up on different sides in the battle between the Presentists and the Eternalists. Presentists maintain that only current objects and experiences have any reality in the philosophical and ontological sense. If a person sits typing at a computer keyboard, that person's body, the keyboard, and the computer are present realities. They are more vivid, more immediately experiential, than any memories the person may have of working at the computer last month. The computer operator's present experiences are also more vivid and have much more impact on that person's consciousness than any future plans to use the computer again tomorrow. Eternalists, on the other hand, argue that there are no discernible differences in the ontological qualities of the past, present, and future. Eternalists are adherents of what is known as the block universe theory.

Scientists and philosophers find it difficult to agree on the ontological differences — if any — separating the past, the present, and the future. What exactly do they mean when they discuss the ontology of time? Our modern word *ontology* is derived from the Greek *ontos*, meaning "to be," and *logos*, meaning "study." Ontology is often regarded, therefore, as the most important component of metaphysics, as well as being inseparable from time-science and from the philosophy of time. Ontology may be defined as the fundamental study of being, of existence. Ontology asks whether there are different levels and degrees of being, and whether one thing's being can actually exceed the being of another. Mass can vary, energy can vary, time can vary in the circumstances that induce time dilation — but can being itself vary? Are we here confronted by an absolute? We assume a thing either exists or it doesn't — but could there be infinite gradations along the ontological spectrum? This is curiously reminiscent of the time argument concerning instantons. Is there an incredibly small and indivisible unit of teleological duration? Or is time infinitely divisible, so that no such thing as an instanton can exist? All studies of ontology — from whatever metaphysical, philosophical, or scientific perspective they evolve — are confronted by the same inescapable fundamental question: *What exists?*

Time-science and time-philosophy are also concerned with the problems of tensed, or tenseless, theories of time. The Eternalists with their block universe theory support the concept of tenselessness. This implies an avoidance of past and future tenses in both our subjective thought processes and in teleological realities. If, during the World Cup soccer games, an observer uses a past tense and says, "Italy has won the trophy," it can be argued by a block universe theorist that the same truth can be expressed tenselessly by using the syntax "Italy wins [or "Italy does win"] the World Cup at time *x*." George Orwell (real name Eric Blair, 1903–1950) used a similar concept in his novel *1984*. His totalitarian regime aimed to suppress language as one means of establishing political control, arguing that if revolutionary words did not exist, then no one could think revolutionary thoughts. Eternalists and devotees of the block universe theory seem to attempt to eliminate tenses in a rather similar way — although there is nothing sinister in their motives.

Another scientific perspective on time is to regard it as what is termed a *fundamental* quantity. The parallel with chemistry is the difference between elements and compounds. Just as a chemical element cannot be analyzed or broken down into constituent parts, neither can

Clepsydra (Egyptian water clock).

time be broken down into other more basic things. It is impossible to define time by using any other quantity — simply because in our present state of knowledge there is nothing more fundamental than time. Mass and space are also regarded as fundamental quantities because it is possible to measure them — just as it's possible to measure time — but it is not possible to break them down into anything more fundamental within the boundaries of our twenty-first-century science. In short, fundamental quantities can be measured but cannot be analyzed, explained, or defined in terms of anything more elemental.

Following this idea of time as a fundamental quantity, it is interesting to note the types of measurements and measuring devices that have been employed over the centuries to try to improve the accuracy of the measurement of time. Returning again to the controversy over the existence of the instanton, vast improvements have been made in the various horological devices with which we measure time with increasing precision. Sundials were among the earliest measurement devices, first appearing around 3500 B.C.

Some two millennia later the earliest known water clocks, the clepsydra, were built in Egypt. Most were designed with two cylinders of water at different heights. As water from the higher cylinder came steadily down a tube into the lower one, the rising and falling of the water levels could be read off against time marks on the cylinders. Later models developed by Greek craftsmen connected the float in the lower water chamber to a vertical rod containing gearwheel-type teeth. This in turn was connected to a cogwheel that rotated the pointer on the face of the clepsydra. Water clocks had an advantage over sundials in that they worked independently of the weather and they also functioned at night. Another elementary timing device was the hourglass, through which a quantity of sand trickled in approximately one hour.

There seems to be an interesting positive correlation between the development of increasingly accurate techniques for measuring time and the general progress of science and technology. Of course, it may be argued easily enough that improvements in time-measuring techniques were simply an integral part of the overall tide of scientific and technological advances taking place over the centuries. However, it may also be conjectured that improvements in the measurement of time contributed certain causative inputs as far as science and technology were concerned.

From approximately 500 to 1300 A.D. there was very little change in methods of time measurement. The aesthetic design and general appearance of sundials altered, but their scientific principles did not. Then, during the fourteenth century, one or two big mechanical clocks in towers were constructed in leading Italian cities. They worked on what horologists call the verge and foliot control mechanism. This is made from a shaft (known as the verge) and a crossbar (the foliot) with a weight at each end. The weights are adjustable, and can be placed at different points along the crossbar. This makes use of the principle of moments of force, which are derived from the force-applied multiplied by the distance-from-the-pivot calculations. A large parent having fun with a small child on a see-saw in a playground sits closer to the pivot than the child does, so that their moments of force are equal and the see-saw is balanced: parent's force-times-distance equals child's force-times-distance. This type of clock control mechanism is referred to as an "escapement" because a certain amount of energy escapes each time one tooth on the gears moves: energy that was stored as potential energy in the weight that drives the clock after that weight has been hoisted to its highest point. In some of the earliest clock towers, these weights could be as much as half a tonne (500 kilograms). Energy can neither be created nor destroyed in normal circumstances — it can only change form.

At the start of the sixteenth century, Peter Henlein of Nuremberg invented spring-powered clocks and watches, which became known as Nuremberg Eggs. Their weakness was a tendency to slow down as the mainspring gradually unwound.

Christian Huygens, a Dutch scientist, was responsible for significant improvements in time measurement in 1656, when he made a pendulum clock with a natural period of oscillation. Huygens's new clock was reliable to within 1 minute in 24 hours, an impressive degree of accuracy (1 part in 1,440). Not satisfied with his first achievements, however, Huygens refined and improved the mechanism until he had achieved an even more impressive degree of accuracy: 10 seconds in 24 hours (1 part in 8,640). In 1671, Bill Clements, working in London, was making clocks with what horologists term the anchor or recoil type escapement. Its advantage over the verge and foliot system was its reduced interference with the pendulum's movements. By 1675 Huygens had produced a balance wheel and spring arrangement, so effective that its principle is still found in some contemporary watches. George Graham found a way to improve pendulum accuracy in the early century by countering the length variations that occurred with temperature changes. His clocks were accurate to within 1 second in every 24 hours (1 part in 86,400). Forty years later John Harrison's marine chronometer reached a degree of accuracy of one-fifth of a second in 24 hours (1 part in 432,000). By the close of the nineteenth century, Siegmund Riefler's clock with its almost completely free pendulum got the accuracy up to one-hundredth

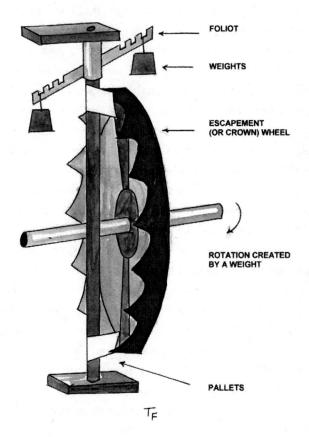

Foliot movement in early clocks.

of a second in 24 hours, and in the 1920s W.H. Shortt produced an amazingly accurate two-pendulum clock, which soon became the astronomers' best ally in many leading observatories. When we place the development of these ever-improving time-measuring devices alongside a timeline of general scientific and technological advances, their similarity is significant.

The quartz clock was the next important development in scientific timekeeping. Totally different from sundials, clepsydra, and clockwork with escapements, a quartz clock functions by using a quartz crystal (silicon dioxide, SiO_2) as an electronic oscillator. The oscillator sends out a signal with a precise frequency, and this signal is displayed as time on the face of the quartz clock or watch. In the right circumstances, quartz will change shape when it's in an electric field. When the field is withdrawn, the quartz resumes its original shape and creates an electrical field as it does so. This field is called piezoelectricity, which can best be defined as a crystal's ability to generate an electrical voltage when the crystal is experiencing some sort of physical tension, stress, or pressure. The process can be thought of as working in both directions: an applied force creates voltage, and an applied voltage creates force. Quartz is ideal in this timekeeping situation first because it can be affected by a simple electrical signal, and second because it changes size so very little with changes in temperature that the difference is insignificant. This also contributes to accurate timekeeping. Contemporary quartz clocks resonate at 32,768 Hertz, which is 2^{15} Hz. This means that they can attain an accuracy of within half a second per day. Sophisticated modifications — such as working the quartz clock along with a computer — make it possible for the clock to measure its own temperature and self-adjust accordingly.

One of the most accurate timekeepers so far developed has to be the cesium clock. The basic principle of such atomic clocks is that the faster they resonate, the more accurate they are. The inaccurate old sundial was based on one resonance every twenty-four hours; the pendulum clock usually worked on the resonance of one tick per second; and the quartz clock produced approximately ten thousand oscillations per second. Since 1967, the cesium clock has been used to define the second (the official SI metric unit of time) with almost incredible accuracy: it is the resonance of the outer electrons of an atom of cesium-133 (9,192,631,770 times a second).

But as if even that wasn't good enough, Hidetoshi Katori from the University of Tokyo has recently created a strontium clock in which the strontium atoms resonate at a rate of 429,228,004,229,952 times each second. This takes us back yet again to the paradoxical nature of our current knowledge of time: we can measure it with great accuracy, but we still have no real idea of what it is that we are measuring. Time is harder to understand than space because of its apparent flow, and yet, unless that perceived flow can be measured against something else — what some physicists have called hypertime — it is impossible to measure. Allowing, for the sake of argument, that hypertime could exist, then, assuming that it also flows, it would have to be measured against a hyperhypertime ... and so on into infinite recession: *reductio ad absurdum*. However, that classical, logical argument may no longer apply to the wilder frontiers of twenty-first-century science. Just because a concept can *seem* ludicrous or absurd to a finite human mind doesn't necessarily mean that it doesn't exist. Werner Karl Heisenberg (1901–1976) — best known for his famous uncertainty principle — was a brilliant German physicist and a Nobel Prize winner. (The Heisenberg uncertainty principle, also known as the indeterminacy principle, states that it is possible to know the position of a particle and its momentum, but it is not possible to ascertain both facts simultaneously.)

Greatly respected as one of the founders of quantum theory, Heisenberg once said, "Not only is the universe stranger than we think — it is stranger than we are *able* to think!" He may well have had the mysteries and secrets of time revolving in his fine mind when he said that.

From these general considerations of the scientific aspects of time, we proceed to the specific contributions of three great time-science theorists: Newton, Einstein, and Hawking.

Chapter 2

NEWTON'S, EINSTEIN'S, AND HAWKING'S THEORIES OF TIME

Nature and nature's laws lay hid in night;
God said "Let Newton be" and all was light.
— "Tribute to Newton" by Alexander Pope

JUST AS IT seemed useful and relevant to begin Chapter 1 by considering what science *is* before investigating the specific branch of science that is concerned with the mystery of time, so it is equally useful and relevant to look at the underlying concepts of matter, energy, motion, space, and time before embarking on the particular contributions made by Newton, Einstein, and Hawking. When time is defined simply as measurable duration or as a continuum without spatial dimensions, so much is left unsaid — and it requires the genius of men like Newton, Einstein, and Hawking to supply at least some of the missing concepts.

One of the strangest and wildest time-mystery theories is the question of whether geniuses like Roger Bacon, Leonardo da Vinci, Galileo, Isaac Newton, and Albert Einstein were astral time travellers, somehow born ahead of their time — twenty-first-century minds occupying earlier physical bodies. The circumstances of Newton's birth were not only strange but tragic.

The authors have a personal interest in Newton because the whole Fanthorpe family (who have now spread all over the world) originated in Lincolnshire, England, just as Isaac Newton did. He was born in Woolsthorpe Manor, part of the tiny village of Woolsthorpe-by-Colsterworth. Baby Isaac arrived prematurely, and was so tiny that his mother, Hannah Newton (née Ayscough), said that he could be fitted inside a quart mug. (The old imperial quart liquid measure was two imperial pints, slightly more than a litre. They were known as quarts — short for quarters — because four quarts comprised an imperial gallon.)

Newton's father, also named Isaac, had died in the British Civil War when his young wife was barely six months pregnant with their son-to-be, who would in due course become the outstanding scientist of his generation. According to the old Julian calendar that was in use at the time, Isaac was born on December 25, 1642. (By the current Gregorian calendar, it was January 4, 1643.)

For those researchers who are particularly interested in the possibility of reincarnation, Galileo had died on January 8, 1642 — not long before Newton was conceived! A vital part of the scientific revolution, Galileo had also been a pioneering physicist, astronomer, and philosopher.

When tiny, premature baby Isaac had survived against all expectations to the age of two, Hannah remarried, leaving Isaac to be brought up by her mother, Margery Ayscough. From local village schools, Isaac went on to King's Grammar School in Grantham, where his budding genius began to reveal itself: Isaac was soon the school's top boy. While

Isaac Newton.

studying there, he lived at the home of William Clarke, an apothecary, and became engaged to Anne — or Katherine in some accounts — Storer, who was William's stepdaughter. As Isaac's studies took more and more of his attention, their relationship faltered. Anne eventually married someone else — a man named Vincent — but Isaac retained the warmest memories of her all his life, and, as far as is known, never had another girlfriend and never married. What an amazing potential genetic heritage was lost there!

Isaac attended Trinity College, Cambridge, from 1661 onwards, and made advanced discoveries in mathematics, including the binomial theorem and calculus. He was also intrigued by the work of Kepler, Copernicus, and Galileo. The scientific aspects of optics and gravitation were other parts of his research. In 1669, Isaac became a Fellow of the college. One of his most memorable scientific achievements was his concept of gravitational attraction, and its dependence upon the mass of the objects that were attracting each other and their distance apart. His law of universal gravitation stated that if the mass of one of the objects doubled, the force between them also doubled, but if the distance between them doubled, the force fell to one-quarter of its previous rating because gravity is inversely proportional to distance.

There were strange mysteries in Newton's life as well as the brilliant science for which he is so rightly remembered and honoured. A murky and misty area of the Rennes-le-Château mystery, which we have investigated in depth since 1975, suggests that Newton was one of the Grand Masters of the enigmatic Priory of Sion. There are numerous theories about the Priory. It may have been an exceptionally old and powerful secret society; it may have been a rival to the Templars in the thirteenth century; it may have been merely a twentieth-century hoax. Whatever it was, there are some controversial and suspect documents connected with the Rennes mystery that list Newton as one of the Grand Masters of the Priory. He is also said to have been a Rosicrucian and to have studied the so-called Bible Code in the belief that there were secret messages hidden within the biblical texts.

On March 20, 1727, Newton died; he lies buried in Westminster Abbey in London. He accomplished a great deal during his time on Earth, but what did he himself believe that time actually was?

Newton was at one period the student of Isaac Barrow, who maintained that there was not necessarily any nexus between time and change. Barrow argued that time exists independently, and that it had existed even before the creation of the universe. Newton agreed with Barrow, and even went a stage further. He argued that time and space together form a container of infinite capacity for all events — and that this limitless container exists independently of whether there are any events in it or not. In an attempt to explain the real nature of space and time, Newton said that although they were similar in some respects to material substances, they were *not* material substances. In Newton's opinion — and it is necessary to remember that Newton was a devoutly religious seventeenth-century man — time and space were independent of matter, motion, and everything else in the universe: they depended solely upon God.

Newton's theories of time distinguished between what he thought of as absolute time and apparent, or relative, time. Apparent time, in Newton's mind, was measured by terrestrial clocks as well as by what seemed to be the motions of the "fixed" stars. He was especially interested in the laws of mathematics, mechanics, matter, and motion — and absolute time made the laws he was seeking that much simpler.

Albert Einstein was born on March 14, 1879, in Ulm in Würtemberg, Germany. His family moved to Munich a few weeks later. As the years passed, Albert continued his education in Switzerland, and he trained in Zurich as a teacher of math and physics. He died on April 18, 1955, at the age of seventy-six.

His father, Hermann Einstein, began his professional life as a salesman, and then ran his own electrochemical factory. Albert's mother had been Pauline Koch before marrying Hermann at Stuttgart-Bad Cannstatt. Although Jewish by birth, Albert's parents were not practising members of the faith, and he actually attended a Roman Catholic school. It was Pauline who insisted on her son having violin lessons, which Albert disliked at the time, although it became part of his legend in later life when he developed a liking for some of the violin sonatas that Mozart had composed.

Events in our early lives can have major effects on our later patterns of thought. Hermann showed his son a magnetic compass when Albert was only five, and, young as he was, the boy realized that something external was making things happen. He understood that something in what he had formerly thought of as empty space was causing the pointer's movements.

As a boy, he loved to make models, and his mathematics ability was always well ahead of his chronological age. Another major event in young Albert's life occurred when his friend Max Talmey, a medical student, and two of Albert's uncles put advanced books on science, math, and philosophy into his eager young hands. One of these was Immanuel Kant's *Critique of Pure Reason*, and it had a profound effect on Albert's rapidly developing mental abilities.

In 1894, Hermann's business failed, and the family moved to Pavia, an Italian city not far from Milan. Albert was already writing his first scientific treatise on ether in magnetic fields, and when only sixteen he carried out the famous thought experiment, referred to in later years as "Einstein's Mirror." He looked into a mirror and then worked out what would happen to his reflection if it could travel at the speed of light. It occurred to him then that the speed of light is independent of the observer. That concept became a central aspect of his theory of special relativity.

His theory of relativity enabled him to make very accurate predictions about the solar eclipse of 1919 because he calculated that light rays from distant stars would be deflected by the sun's gravity. When his predictions proved correct, he became world famous — a legend in his own lifetime and a synonym for genius.

Einstein's magnificent mind was at least the equal of Newton's, but Einstein had the additional ability to see around the corners of time and space and conceptualize them in a way that Newton hadn't done. Einstein also perceived what he regarded as two essentially independent theoretical systems. One was quantum theory. This deals with the ways in which tiny components of the universe behave — things like atoms and subatomic particles such as electrons — and is also concerned with what properties they have. The earliest version of quantum theory set out to explain how electrons remain in their orbits; many of its contemporary theories couldn't do that satisfactorily.

The study of electromagnetic waves — of which light is a prime example — led Max Planck in 1900 to describe the energy in such waves as small packets that he

Albert Einstein.

called quanta. (In Latin, the term *quantum* means "how much.") Einstein used Planck's concepts to demonstrate that wave energy could also be understood in terms of a particle (such as a photon) with a fixed amount of energy related to its frequency. From this came the theory known as wave-particle duality: the idea that wave-particles were neither waves nor particles but had some properties that were applicable to both!

Einstein's second theory was relativity. Strictly speaking, there were two theories of relativity: general relativity and special relativity. The famous Michelson-Morley experiments revealed that the velocity of light was absolute. It didn't make any difference whether the speed of light was measured with equipment moving quickly towards the light source or equally quickly away from the light source: the velocity of light remained constant at 300,000 kilometres per second (186,000 miles per second).

One of the curious enigmas of relativity is that two observers who are moving relative to each other will experience both time differences and length differences. If each observer has a clock and a yardstick, each will experience the other's calibrated rule as shorter than her own, and the observers's clock will be going faster than the observed clock!

Einstein realized that quantum theory and relativity were not *necessarily* contradictory, but neither did they seem capable of being fused into one unified theory, which was what Einstein wanted — in very much the same way that Newton had wanted a simplified, absolute time that would make his laws of mechanics easier to understand and easier to handle.

As recently as 1954, Einstein said that physics had no general, theoretical basis, no logical foundation. He went on to argue that if what might be called the axiomatic basis of physics cannot be extracted from experience and observation, and we have to try to invent it, might it ever be possible to find the objective truth behind everything — if such a truth indeed exists?

Always a positive, optimistic character, Einstein expressed his confident opinion that there *was* such an objective truth behind the universe, and that the human mind would be capable of finding it eventually. He believed fervently that pure thought was capable of grasping reality.

For Newton time was absolute. For Einstein it was relative, and capable of dilation in certain circumstances. The next brilliant thinker to examine the intriguing question of time was the outstanding relativistic cosmologist and mathematician Stephen Hawking. Newton's theories worked satisfactorily in weak gravity. Einstein said that warped time and curved space enabled gravity to be described: his ideas — unlike Newton's — worked well in strong gravitational fields. But Einstein's relativity yields to quantum mechanics when we consider singularities like the Big Bang and the curious things that happen in black holes. It is Hawking's genius that is capable of welding quantum mechanics and relativity into a genuine unified field theory.

It is important for an understanding of Hawking's work and its relevance for time-science to remember that as far as is known to contemporary science there are four forces in the universe. First comes gravitational force, which is responsible for planets, stars, galaxies, and nebulae. Second is the electromagnetic force, which is the foundation of all chemical reactions and which can be thought of as the force that keeps atoms together. Third is the strong nuclear force, which holds protons and neutrons inside the nucleus of an atom and which is an integral part of nuclear fusion and nuclear fission. Fourth is the weak nuclear force, responsible for the radioactive decay that occurs when alpha and beta particles "leak" out of an atomic nucleus spontaneously. Cosmologists like Hawking believe that these four mysterious forces became distinct from one another in the earliest moments of the universe's existence. Was time there before them, or did it come into being with them?

The recent work of Hawking and other daringly gifted relativistic cosmologists has pointed to the different cosmological models that relativity can encompass. Some of these models are able to start from a singularity, expand to a given size, and then contract again. If time is operating within such a model, does it start to run backwards as the model contracts? A second type of cosmological model expands at different rates in different directions. A third type expands forever — there is no destructive return to a singularity for them, they simply go on reaching one magnitude after another. All these very different models are nevertheless compatible with Einstein's relativity equations. Hawking's contribution here has been to work with his friend and colleague Jim Hartle to assign a wave function to each of the three main types of cosmological models. In theory, that wave function will indicate which way the model will behave: expansion-contraction; uneven, irregular expansion; or smooth, infinite expansion. Working on models of universes that have no boundaries of space or time, Hawking and Hartle have calculated that such no-boundary models are not dissimilar from observations made in this actual universe in which we find ourselves.

We are all deeply indebted to Newton, Einstein, and Hawking. Those three great scientists have made gigantic contributions to our knowledge of space and time — but they still leave some profound and fundamental questions unanswered. What *is* time? Did it have a beginning? Will it have an end? Is there any way in which human consciousness can either escape from its tyranny or learn to control it? Is time cyclic or linear? Is it finite or infinite?

Chapter 3

CYCLIC OR LINEAR TIME — FINITE OR INFINITE

THE SCIENCE of time, and the significant contributions made to it by Newton, Einstein, and Hawking, lead on to what might best be termed the metaphysics of time: a consideration of whether time can best be thought of as cyclic or linear, as finite or infinite.

Working with his friend and colleague Thomas Hertog, the brilliant Stephen Hawking has just proposed a new theory about the beginning of the universe. Hawking and Hertog consider the possibility that there need not be an either-or situation as far as the history of the universe is concerned — there could have been several histories. There was not necessarily just one unique starting point and one unique formation and development process: the universe could have experienced all of the alternatives soon after the Big Bang. According to the Hawking and Hertog hypothesis, what we have now is, in a sense, the surviving and persistent history of the universe that leads to our present *now*. String theory postulates the idea of many possible universes, some with different space-time dimensions from what we think we are experiencing here. Hertog and Hawking's new work suggests that these alternative universes simply faded relatively soon after the Big Bang. What possible combinations of space and time faded with them? Do they still exist *somewhere* as probability tracks, as the Worlds of If?

What size is time and what shape is it — and do such questions have any relevance and meaning? Is time so strange and so incomprehensible that it eludes human thought?

John William Dunne was born in 1875 and worked as an aeronautical engineer, dying in 1949. Although he was practical and pragmatic, Dunne had an inexplicable precognitive dream in which he saw the eruption of Mount Pelée in Martinique in 1902 — or else he read an account of it in a newspaper that had not yet been printed! This was the beginning of his investigations into the nature of time, and led to his theory that during dreams our human experience of time can be non-linear.

His book *An Experiment with Time* was first published in 1927. In it he argues that the philosophical problem of infinite regress confronts any thinker who is trying to solve the mysteries associated with our concepts of self-consciousness, volition, and time. Dunne expands these basic concepts by saying that a self-conscious entity is a being that knows that it knows; an organism that wills is one that can examine alternatives and make choices based on its motives; and time is a series insofar as it is necessary to be in another time in order to observe the first one, and yet a third time to observe the second. This, according to Dunne, leads to an infinite regression, or infinite recession — a series of times that go on forever.

But such concepts as infinite regression are unpopular with serious thinkers and philosophers who aim to find effective solutions to problems, not merely play with erudite words. At the start of his work, Dunne set out to eradicate the apparent infinite regress of time, which had puzzled him since childhood. He even had vivid memories of asking his nurse about it —the type of question that intelligent children can ask and that few, if any, of the most sympathetic and helpful adults can ever answer. As Dunne himself says in his books (he also wrote *The Serial Universe* and several

Brilliant twentieth-century time theorist J.W. Dunne.

others) he did everything he could think of to dispose of the ideas of the infinite regression of time, of self-consciousness, and of the will. Infinite regression withstood all the attacks he could launch against it. He came to the conclusion that far from dismissing the concept of infinite regression, he should regard it as logical and valid — the true foundation of all epistemology (the theory of knowledge).

Dunne's theory of infinite recession.

As Dunne's thinking along these lines developed, he linked the philosophical concept of regress to the mathematical one. High-school math examiners love to confront students with a series of numbers and then invite them to find the next two terms in that series, along with the formula that generates the series. Suppose that a simple series in such a math question runs: 5, 9, 13, 17, 21 ...

The difference between the terms is 4 in every case. The letter n refers to the number of the term in the sequence. When $n = 1$ the nth term is 5; when $n = 2$ the nth term is 9; when $n = 3$ the nth term is 13; and so on. As the difference between the successive terms in this series is always 4, it is referred to as a common difference series. The formula for such a series is: nth term $= dn + (a - d)$, where $a =$ the first term in the series (which in this example is 5) and $d =$ the common difference (which in this example is 4). The formula for the nth term for this series then becomes: nth term $= 4n + (5 - 4)$, or nth term $= 4n + 1$.

Assuming that we want to find the nth term when $n = 6$, we substitute 6 for n in the formula and get: $(4 \times 6) + 1 = 25$. This is clearly correct because adding 4 to the fifth term (21) gives the same answer. Suppose, however, that we wanted to find the 123rd term (where $n = 123$). It would be laborious and time-consuming to go on adding 4s to the terms already given in the series. The formula shows that $(4 \times 123) + 1 = 493$.

Other mathematical series — including those that do not have common differences — are equally susceptible to nth term formulae, and these formulae can take the series to infinity. Dunne argues that a regress is a mathematical series, and that such a series is the expression of a relationship, like the nth term formula in the previous example. He also argues that it is not possible to figure out such relationships if we have only the first term in the series: we must have several terms before we can work out the nth term formula governing the series. Applying these arguments to time, Dunne maintains that humanity has failed to solve the mystery because we have confined ourselves to the first term in the time series. Progress can be made only after we start to look carefully at the other terms in the time regression.

The practical side of Dunne's work on time theory was the careful recording of dreams to see if they were accurately precognitive, and he used both his own experiences and those of others who reported their apparently precognitive dreams to him. A major complication in all of Dunne's work in this field was that "future" events in his serial universe theory were not finite and unavoidable, but could be altered.

A significant suggestion from Dunne's work is that we do make use of fragments of possible futures when we "construct" our dreams, as well as when we express current anxieties and emotions and past memories in them. If these glimpses of possible futures — probability tracks that may or may not become actual future experiences — are real, then it looks as if the dreaming self is not restricted to the present moment. Based on all Dunne's research, there is nothing anomalous, paranormal, or supernatural about the time-free activity of this dreaming self: all of us do it continually to a greater or lesser extent.

Dunne's arguments can be summed up by his statement to the effect that if time is a serial phenomenon, then the universe itself has to be serial. He has never gone unchallenged, but he was a worthwhile, pioneering thinker whose contributions to the ongoing debate about the real nature of time deserve to be treated with respect.

How many precognitive dream experiences do we need to support Dunne's intriguing theory? As well as his dream forecast of the Mount Pelée volcanic eruption, Dunne had a vivid precognitive dream of a railway crash involving an embankment before the Flying Scotsman crashed near the Forth Bridge.

The biblical Joseph not only experienced dreams of his own future greatness, he also interpreted the dreams of others: but there are curious little inaccuracies and anomalies in his dreams. In his dream that the sun and moon would bow down to him, the two celestial bodies represented his parents — yet Jacob's beloved wife Rachel (Joseph's mother, the moon symbol) was already dead. Joseph did correctly interpret the dreams of Pharaoh's butler and baker — the former was restored to royal favour, the latter was executed. Pharaoh's own dreams about the ears of corn and the cattle were also interpreted accurately as years of plenty followed by years of famine.

In these examples it seems that Dunne's ideas about probability are supported: the main gist of a future event seems to be there, but a few details are wrong. The dreams showed what could have been, what should have been. If Rachel had not died young, she too would have bowed to her illustrious son, the Viceroy of Egypt. Another biblical dream refers to Gideon's victory over the Midianites. (Judges 7:13–14) One of Gideon's soldiers dreamt that a cake of barley bread tumbled into the Midianite camp and knocked a tent down. Gideon, the courageous and skilful military leader, was a miller in civilian life. The barley bread symbolized the man who made the flour from which it was baked. Another biblical dream account tells how King Nebuchadnezzar had a nightmare that he wished to have interpreted. Daniel told him not only *what* he had dreamt but what the dream *meant*. Nebuchadnezzar had dreamt of a huge idol made from gold, silver, brass, and iron. Its feet were iron intermingled with clay. A stone was cut out without hands, and it broke the feet of the huge idol, which was smashed to pieces. This stone later grew into a gigantic mountain that filled the whole Earth. Daniel explained to the king that the stone that filled the whole world represented God's eternal kingdom. Another part of the biblical account tells how Daniel warned the king that unless he complied with God's law, he would suffer dire consequences. Nebuchadnezzar became mentally ill and acted like an animal eating grass in the fields. Only when the king repented was he healed and his kingdom was restored to him.

In the Old Testament account of King Solomon's dream, God asked him what gift he would like, and Solomon chose wisdom so that he could rule fairly and justly for the good of his people, not his own selfish aggrandizement. His request was duly granted, and many other benefits were added to it.

The Gospel of Matthew in the New Testament records a number of dreams connected with Joseph, Mary's husband. The first reassures him that her premarital pregnancy is divine in origin; another warns him to escape to Egypt because of Herod; a third tells him that it is safe to return as Herod is now dead. Matthew also records the dreams of Pilate's wife as a result of which she begged her husband to have nothing to do with the priestly plot against Jesus that led to his crucifixion. (Matthew 27:19.)

During the fifteenth century, in the small market town of Swaffham in Norfolk, United Kingdom, where co-author Lionel was a student at Hamond's Grammar School in the 1940s, there lived a peddler named

John Chapman. John had a very vivid recurring dream in which he was told that if he went to London and stood upon London Bridge he would receive information that would make him rich and famous. After spending many hours standing forlornly on the bridge, John was approached by a local resident who suspected that he might be an opportunistic thief, watching to see when people left their houses. John protested his innocence and explained that he had come to London because of the strange dream.

"If I were such a fool as you," laughed the Londoner, "I'd be on my way to a place I've never heard of called Swaffham, where in my dream there was a crock of gold buried under an apple tree." John wisely kept his own counsel, chatted inconsequentially for a few minutes more, and then set out for home. Arriving late at night, he asked his wife to fetch a lantern and spade. More than a little surprised, but very glad to see her husband safely home again, she held the lantern while John began digging under their apple tree. Not far down, the spade rang out on a sturdy earthenware pot, which turned out to be filled with gold coins.

Keeping their newfound wealth a closely guarded secret, the Chapmans cleaned the crock, which bore a strange inscription they could not read, and then set it in the window of their cottage. Before long, as they had half-hoped when they cleaned up the jar and put it on display, two students from Cambridge who were travelling through Swaffham on their way home at the end of term knocked at the door.

"You have a very interesting old jar there," said the students politely. "May we ask where you found it?"

"A long way from here," said John cautiously. "My travels as a peddler take me far and wide. I can't remember exactly where it came from — but it was many miles from here."

"That's a pity," said the students, "because this ancient Latin inscription says *sub mihi illic est a maioribus.*"

"What does the Latin mean?" asked John.

"Under me there is a greater," answered the students.

John and his wife gave them refreshments, and saw them safely on their way. Then, as soon as the coast was clear, John started to dig again. The old Latin inscription was true. A little farther down, the Chapmans found a far bigger earthenware container, crammed with gold coins.

Having known what hard work and poverty were like, John and his wife gave generously to the poor and were significant benefactors of Swaffham Church.

From 1435 until 1474, the Reverend John Botewright was rector of Swaffham. He carefully recorded all the work that was done on Swaffham Church during his incumbency. His records reveal that John Chapman paid for the rebuilding of the north aisle, and one carved pew-end still shows a peddler carrying his pack accompanied by his faithful dog.

John Bellingham assassinated British Prime Minister Spencer Percival on May 11, 1812. Many details of the tragedy featured in a Cornishman's dream eight days before the Prime Minister was fatally shot.

Not all precognitive dreams are as strange and traumatic as the one that apparently foretold Spencer Percival's assassination. Some are so ordinary and matter-of-fact that they are rarely noted down simply because of their ordinariness. One of these cases, however, was noted by Charles Dickens, whose powerful references to the paranormal in *A Christmas Carol* indicate his interest in the subject. Dickens had a vivid dream about a lady in a red shawl, to whom he was introduced. Her name, Miss Napier, was given to Dickens very clearly in the dream. He knew of no one by that name, and the features of the lady in the dream were not familiar to him. The next day, however, two of Dickens's friends, Miss Boyle and her brother, called to see him. They brought a friend with them — a lady wearing a red shawl, whom they presented to Dickens as Miss Napier.

The Maria Marten red barn murder case involved what *may* have been a prophetic or revelatory dream on the part of the dead girl's stepmother. It is also possible that the alleged dream might have been only an elaborate component of a plot to destroy William Corder, who was hanged for Maria's murder in 1828. We will assume for the sake of argument that the dream was genuine and spontaneous. The outline of the case is that Maria vanished in 1827, having allegedly married wealthy farmer William "Foxy" Corder and gone away with him from their home village of Polstead in Suffolk, England. Following Maria's long, unexplained failure to communicate with her family, her stepmother claimed that she had dreamt more than once that Maria had been murdered and that her body lay concealed below the floor of the red barn on Corder's farm. She finally persuaded her husband to search the portion of the barn where the dream had indicated Maria's body lay — and the girl's badly decomposed remains were duly exhumed. Corder, who had married another woman only a few months previously, was running a girls' school in London with his new wife. He was arrested and brought back to Bury St. Edmunds to be imprisoned, tried, and executed.

Co-author Lionel's mother, Greta, was born on June 28, 1914. On that very same day Archduke Franz Ferdinand of the Austro-Hungarian Empire was assassinated in Sarajevo: an assassination that triggered the First World War. On the night before the assassination, Joseph de Lanyi — who was bishop of Grosswarden, and had been Franz Ferdinand's tutor — had a terrifying nightmare during which he dreamt that he had received a black-edged letter with the archduke's heraldry on it. As his nightmare continued, the bishop saw Franz Ferdinand and his wife, Sophie, gunned down by an assassin as they rode in their car. De Lanyi also saw some clearly legible words, as though they were written in the dream letter. Addressed to him, and written in the first person by his former royal student, the letter told how he and Sophie had been shot and killed.

Because it is human nature to want the world to be an exciting, adventurous, and romantic place rather than a dull, mundane, and monotonous one, it is important to consider the possibility that Dunne and his time-transcending dream theories could be wrong. It is helpful to balance Dunne's arguments by using the illustration of some delicate musical instrument — a sophisticated electric organ, for example — that needs to be played continuously if it is to be kept in optimal condition. When the organist (the conscious mind) stops playing his deliberately planned and chosen melodies, the organ's equivalent of an autopilot takes over and produces a mixture of random notes, discords, one or two almost recognizable melodies, and a lot of weird, chaotic noises. If that organ symbolizes the human brain, is that how our dreams are generated? It also has to be recognized that statistically, if enough sleepers dream enough dreams through enough nights, *some* of those dreams will be paralleled in what we cheerfully think of as "the real world" on some future occasion.

Aristotle (384–322 B.C.) was keenly interested in dreams and their meanings. In "On Dreams," he wrote: "The best dream interpreters are those who are skilled at observing similarities. It is easy enough to interpret plain and simple dreams: anyone can do that. But some dream presentations are like reflections in moving water — and it is very difficult to guess what the original is."

In addition to the Freudian and Jungian psychological interpretations of dreams, there are numerous interesting old volumes available that list some of the traditional and generally accepted meanings of typical dreams. These folklore dream guides are like code books for Morse code or semaphore, giving the widely accepted meanings of various familiar dream symbols as if they were standardized and accepted ciphers. All of these interpretations are interesting, and some, perhaps, are significant. A few examples make this clear. Dreams of acorns indicate good things and prosperity. To ride a horse or a bicycle uphill in a dream means good things are coming to the dreamer. Candles have mixed meanings in dreams: burning steadily suggests good news; flickering or being extinguished is an omen of failure and loss. To dream of being under a dome indicates a change for the better. Elephants are very good dream symbols — they forecast prosperity and success. When the dreamer walks through fog, she has business, relationship, or career worries. Grass is an ideal dream symbol: it indicates continuing good fortune and success in business. Harvest dreams are just as good as dreams of grass: dreaming of a busy harvest scene forecasts prosperity. Dreams of iron have mixed meanings, just like dreams of candles: strong, new iron in a smithy or workshop represents power; rusty iron symbolizes failure and loss of money. Jam is another interesting dream symbol: it is said to symbolize happiness associated with a journey. Dreams of knives foretell quarrels and serious disagreements. A ladder is another good dream symbol, suggesting that the dreamer will have great success in her chosen field. Dreams of mistletoe symbolize happiness and celebrations. Nets are not very positive symbols — they suggest trouble, entrapment, perhaps business or career problems. They can also suggest that the dreamer is a cunning and ruthless character. Orchestral dreams are good, symbolizing harmony in life, many good friends, and family or team successes. Paper is a dream symbol for law courts, litigation, and the futilities of bureaucracy. Dreaming of quick-sand is not good: it foreshadows long delays, procrastination, and problems of various kinds associated with those postponements. Rams are good dream symbols: they indicate loyal and powerful friends who will help and fight for the dreamer. If you dream of scales — like those that symbolize justice — you have a strong sense of fairness and your future actions will be governed by this. A torch is a good and positive dream symbol, signifying successful leadership on the part of the dreamer. It is a particularly favourable omen for sports people, politicians, and professional entertainers. Uniforms, either being worn or on display in a shop window, are also good dream symbols: they suggest effective organization, logical thinking, and the kind of consistent methods that lead to success. Dreams of victory are especially positive symbols — they forecast victories and triumphs in the real world. A wall is another mixed symbol: jumping over it represents overcoming obstacles and solving problems that had blocked the dreamer's path, but being unable to climb it or to find a door indicates frustration and delay. To dream that you are playing a xylophone — or watching one being played — suggests that a great deal of concentrated effort and activity is needed to achieve your aim; objectives will be achieved in a roundabout way. Dreams of yachts are good: they symbolize happiness and enjoyable relaxation. Dreams of the zodiac are among the most favourable of all: they indicate success and happiness in a great many aspects of life.

Working on specific predictive dream analyses, and looking carefully into traditional dream symbols like these, persuaded Dunne of the serial nature of time and of the universe. But there are several other possible explanations of the metaphysical mystery of time and of what may best be called the topography of time.

The first of these is that time is circular — but circularity is not a simple or unitary concept. The perfect and regular one-dimensional curve is simply the circumference or perimeter of a circle. A flat disc is circular, but now the circle has two dimensions that give it area. In three dimensions the basic circle can become a cylinder or a sphere. But can we imagine an extra dimension added to a sphere? The flat disc or cylinder can roll in only two directions — backwards or forwards — as the appropriate force is applied to it. A sphere can roll in any direction when an appropriate force acts upon it. What if a hypersphere existed — one with an extra dimension over and above the three dimensions of a sphere? Could such a hypersphere move through time? Or, more significantly, could time itself be a hypersphere? Are we, who experience time as a unidirectional flow, actually existing on the super-dimensional surface of a hypersphere?

If time has the property of circularity — any of the four conjectural circularities considered here — can it "roll" in such a way that events seem to us, as observers, to occur more than once? Is such a hyperspherical time hypothesis compatible with Dunne's serial universe theories? Is it possible that one hypersphere contains another, which contains another, which contains yet another ... rather like those nesting Russian dolls? The circularity of time is not incompatible with the seriality of time, and neither concept is incompatible with the data that Dunne collected on precognitive dreams. If human experience of the time arrow phenomenon is envisaged as teleological movement from an inner concentric hypersphere to one that is topographically similar but larger and further out, then the observer's experiences at that higher level could well be confusingly similar to those already experienced on the level within or below. It could also be that these concentric hyperspheres have a singularity at their centre, and that singularity might be the same phenomenon that relativistic cosmologists like Hawking and Hartle regard as the Big Bang — the point from which time itself may have begun.

Frank Tipler, professor of mathematical physics at Tulane University in New Orleans, is a truly remarkable and fearlessly independent thinker. His challenging theories on what he terms the "omega point" are explained in his book *The Physics of Immortality*, in which he predicts that the evolution of intelligent entities — such as human beings — will enable science to progress exponentially at a speed that overtakes the collapse-rate of the universe. Tipler then argues that this can provide infinite "virtual time" during which there will be unbounded possibilities, including computer simulations of all intelligent beings who have ever lived during the universe's entire existence: another way, in his view, of looking at the meaning of resurrection and potential immortality. But although Tipler is known mainly for his omega point theory, it is his Tipler cylinder that is significant at this point in our consideration of the circularity of time. The key point in the Tipler cylinder theory is that it *revolves* — and the most useful and understandable model is the phenomenon of the gyroscope.

The Tipler cylinder theory in summary is that if a cylinder with enormous mass — equivalent to the combined mass of several neutron stars — was made to revolve rapidly around its longitudinal axis, it would induce frame-dragging effects that would in turn warp space-time. The effects would become most powerful as the rotating Tipler cylinder approached the speed of light. Theoretically, these frame-dragging phenomena would tilt what are called the "light cones" of any objects that were close enough to be affected by the enor-

The gyroscope demonstrates the principle of the Tipler cylinder time machine.

mous mass of the rapidly spinning cylinder. If the theory is correct, then parts of the tilted light cones will point backwards along the time axis. This could, hypothetically, make time travel possible, which will be examined in greater depth in Chapter 4. In effect, a Tipler cylinder would appear to play games with time in much the same way that a rapidly spinning gyroscope appears to play games with gravity.

There is also the question of whether time is linear, rather than serial or circular. All our sociological and psychological experiences of time prompt the common-sense view that time *is* linear, and that it also has a unidirectional flow. Centuries of observation, however, have taught us not to be overconfident when it comes to accepting common-sense explanations of the universe and our place in it. Yesterday's certainty is tomorrow's modification. The Earth is not flat. The Earth is not the centre of the universe. When you stand at the North Pole, the only direction in which you can travel is south; and when you stand at the South

Pole, the only direction in which you can travel is north. Very high gravity and very high speed can cause time dilation.

Conversely, common sense is not invariably wrong. Fresh, organic fruits and vegetables, vitamin supplements, adequate sleep, and regular exercise will increase health, fitness, and longevity. Cigarette smoking, drugs, obesity, and too much alcohol will wreck the human body and shorten its life span significantly. Stress, anxiety, worry, and other forms of maladjustment to life's problems can induce mental illness. Applied common sense, tranquility, relaxation, optimism, confidence, and a sincere, loving, kind, gentle, and positive attitude towards others will establish and reinforce mental health. Sometimes truth is obvious and straightforward. But *can* time really be what our senses and the majority of our experiences of life tell us that it is?

Is time linear? Is it simply a unidirectional flow? Is it a one-way street leading from the past through the elusive, fleeting present and on from there to an uncertain and as yet unformed future? Is time connected mysteriously with 3-D space as Einstein tells us when he refers to a space-time continuum? Is it infinitely regressive and serial? Is the stark truth that we are the hapless victims of time: we are born, we mature, we age, and we die? This is the point at which science, metaphysics, and philosophy take over from simple common sense.

Common sense tells us that if we hit a golf ball hard enough it will fly several hundred yards, but if we teed up an egg and struck it in the same way it would merely shatter messily. What common sense fails to tell us about either of these events is that the golf club, the muscles that swing it, the golf ball that flies, and the egg that shatters are all made up of extremely small particles called atoms, and that these in turn consist of a nucleus full of protons and neutrons surrounded by relatively large volumes of empty space in which indescribably tiny electrons move in different "shells" or energy levels. Common sense cannot tell us that these electrons are unstable and volatile unless they have eight in their outermost shell, so that sodium, which has only one, rushes to bond with chlorine, which has seven: a dangerously reactive metal and a hazardously reactive toxic gas unite and turn into ordinary table salt. Common sense has no explanation to offer for the phenomenon of chemical reactants combining to make compounds that have very different properties from those of the original reactants.

It is science that supplies the explanation, tests it, and proves it — no matter how improbable it seems when viewed through the simplistic lens of common sense. Just because time gives the *impression* that it is linear and unidirectional, it doesn't have to be so — and if Newton, Einstein, Hawking, Dunne, and Tipler are our guides, then time is something much more mysterious and much harder to understand than simple linearity.

Is time finite or infinite? The question is almost unanswerable. If one school of thought regards time as starting with the Big Bang and perhaps ending if the universe contracts into a similar singularity, then time can be regarded as finite. But if time existed before the Big Bang, and, in a sense, continues after the terminal singularity has absorbed everything else into itself, then perhaps time is best described as infinite. There is also the intriguing question of the finite or infinite nature of what we call an instanton: the smallest possible division of time. If the instanton exists, and is in itself incredibly small, is there a kind of infinity in its smallness? How wide is the eternal *now*? If an instanton exists — however briefly — *where* does it exist? If an infinite volume of 3-D space simultaneously acknowledges one specific instanton of time, does the infinite space that acknowledges it confer the property of infinity on the instanton itself? And does it then confer infinity on the mysterious phenomenon of time as a whole? If Hertog's and Hawking's latest theories are correct, does infinite time still exist in the so-called faded universes, and how complete is the fading process?

In the next chapter we consider whether time is impregnable to both matter and energy, and, if it isn't, whether time travel is possible.

Chapter 4

TIME VERSUS MATTER AND ENERGY: IS TIME IMPREGNABLE? CAN WE TRAVEL IN TIME?

THE RELATIONSHIP between matter and energy is elegantly summed up in Einstein's famous equation $E = MC^2$, where E is the energy released or available, M is the physical mass, and C is the velocity of light. In Einstein's theory of the space-time continuum, matter tells space how to curve, and space tells matter how to move. In Einstein's world of mathematical physics and relativity, matter and energy may well be interchangeable at that prodigious rate. In the previous chapter we also looked at the phenomenal powers of a Tipler cylinder rotating at close to the speed of light — a vast mass moving with huge rotational energy. If Tipler's theories are correct, such a tremendous combination of mass and energy can affect time — or, more accurately, perhaps it is our human experience of time that is affected, rather than time itself. Whichever way it works, a Tipler cylinder would, in theory, enable observers to go backwards in time. There is, however, a paradoxical problem connected with this type of time

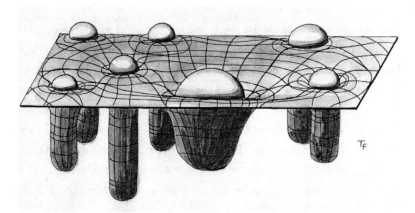

Matter tells space how to bend — and space tells matter how to move.

travel: it seems unlikely that the observer hoping to use it for time travel would be able to go any further back than the point in time at which the Tipler cylinder was first created and used. A historian in the year 2267 who wanted to come back to look at us could only do so if a Tipler cylinder was already up and running in the vicinity of our solar system today. If the Tipler mechanism was completed and started up only in 2267, that's as far back as the hopeful historian could go. That paradox may provide one possible answer to the age-old question: If time travel does eventually become possible, why haven't we already met time travellers from the future? It may be because they cannot come back any further than the point in time when their time machine was invented, built, and started up.

If the Tipler cylinder paradox prevents teleonauts or chrononauts (as we may come to call them one day) from going back to visit the pre-machine past, can they get into the future or move *across* into other probability tracks? Hertog's and Hawking's theories involving the possibility of multiple "faded universes" in the remote past — together with the challenge of string theory — would seem to hint that movement between alternative universes might not be impossible. The space-time systems of such alternative universes might be totally different — not only from ours but from one another's. This was not lost on the superbly creative mind of C.S. Lewis in his Narnia series. When the Earth children were having adventures in Narnia, events that occupied years there took only a matter of hours on Earth because of our very different space-time systems.

String theory is one of the most important ideas available to time-scientists: the problem is that it goes a long, long way beyond their central focus — the nature and behaviour of time. String theory has sometimes been popularly referred to as the "theory of everything," because it sets out to suggest answers to more than one of the vast questions that intrigue scientists, philosophers, theologians, and metaphysicians. What's the universe made of? How did it get here? Who and what are human beings — and why are we here? *Here* in this sense has to refer to our position in time as well as to our 3-D coordinates in space. *Here* has to be a combination of both *when* and *where*.

The so-called standard model of particle physics, which we have already encountered in summary form, also tries to answer some of those ultimate questions, despite its limitations. It provides descriptions of the rudimentary substances or particles comprising the universe, and the forces that cause them to interact. In outline, the standard model talks about twelve fundamental components: half a dozen are quarks with the intriguing names of *strange, charm, top, bottom, up,* and *down,* and a further half-dozen very, very small objects are referred to collectively as *leptons.* Three of these are neutrinos; one is an electron; one is a muon; the last is a tauon.

Again, it is the standard model that recognizes the four fundamental forces of physics: electromagnetism, the weak and strong nuclear forces, and gravity. It is this last fundamental force that causes the physicists' problems. It has been almost impossible so far to work out a quantum theory of gravity, but string theory seems to be the leading contender at the moment, and it goes a lot further than solving the riddle of the quantum theory of gravity. It holds out the possibility of revealing the secret of the universe's fundamental structure.

Like all great scientific ideas, string theory is awesomely simple. Put in its briefest and most rudimentary form, string theory suggests that *all* the fundamental particles — the tiny little bits of substance of which the universe is constructed — have one fundamental truth in common: they are all different arrangements of "string."

If string theory really has answered the riddle of the universe, we have to re-evaluate a minute particle like an electron, for example, not as an indescribably small point but as something more like a small loop of string! These so-called string loops can do many things that fixed points can't. Strings can oscillate — and if they oscillate in a variety of different ways, they will be observed in a variety of different ways: as photons, as electrons, as quarks ... and even as something more mysterious that has yet to be positively and definitively identified.

Another fascinating perspective on string theory is that if it's true, it adds weight to the likelihood that there are probability tracks — which means that time may not be a single, unidirectional line or a circle after all, but an ever branching "tree" of coexistent, probable universes. Maybe the secret of time travel is not to go backwards or forwards — but sideways across an infinite number of probability tracks.

These tracks — if they exist — might hold out optimal and pessimal* life experiences: the best and the worst things that could possibly happen to any individual. In his optimal world, the person enjoys love, health, happiness, wealth, power, safety, and security. In his pessimal world, the person is hated and despised, plagued by chronic and painful illnesses, the deepest misery, destitute poverty, abject weakness, and terrifying insecurity. Imagine portals opening between probability tracks — alternative universes — through which an emaciated, pain-wracked, dying beggar can glide from his pessimal World of If into his optimal World of If, where he is a fabulously wealthy, greatly loved, and respected ruler, enjoying robust health and tireless energy, surrounded by a loving family and honest, sincere friends.

Moving on from probability tracks and the Worlds of If, our next analysis aims to probe the *vulnerability* of time. Let us accept that string theory becomes largely proven and widely accepted, superseding not only Newton's ideas but, to some extent, Einstein's relativity and space-time continuum theories as well. If time in the classical sense is impervious to interference of any kind, from any source, why are there so many well-documented, widely reported, and seemingly authentic cases of time slips, warps, and glitches? Unless *every* case of the abnormal behaviour of time is a deliberate fraud or hoax, or a simple error committed in good faith, then time looks as though it is not totally invulnerable to the powers of mass and energy. If light can be affected by gravity, can gravity also affect time? What about the electromagnetic forces? Is it possible that gamma radiation can penetrate and modify time? Or perhaps it is something subtler and less tangible — like the power of the human mind — that may be responsible for these frequently reported slips, warps and glitches in time? If time does turn out to be vulnerable to one or more external influences, how does that affect the likelihood of time travel?

What forms might time travel actually take? Consider first the type of time machine that H.G. Wells wrote about: literally a machine, a physical device of the same basic generic pattern as a car, a plane, a space shuttle,

---

* Pessimal: the authors' favourite Latin-based antonym for optimal. i.e. Optimal = best possible. Pessimal = worst possible.

a boat, or a submarine — but much more advanced. The second type is what might be termed a time portal. Theoretically, these doorways into time are openings or gateways at the interface of different epochs or probability tracks, through which the chrononaut can pass. Here is no daring pilot directing and controlling a machine as Wells's hero did. This chrononaut is more like a caver or potholer: a speleologist of time, passing from one part of a strange labyrinth of unexplored possibilities into another. It seems highly likely from the evidence of those who report occasional experiences of time slips, time warps, and time distortions that these portals occur naturally for reasons that are not yet even partially understood. But if something akin to a Tipler cylinder could actually be deliberately constructed by a highly advanced technology, or if lightspeed and gravitation could be harnessed and directed towards opening such a portal, then there would be a sense in which those who had designed and made it were in charge of its openings and closings — at least to a limited extent. Here the chrononauts afloat on the rivers and canals of time would be largely dependent upon the scientists and engineers who were opening and closing the teleological equivalent of locks and sluices. The direction of this type of time travel would depend ultimately upon the pattern of the openings and closings of the gateways.

Gaining such control over the interfaces of differing times and probability tracks by scientific and mechanical means may well be a genuine possibility. So our chrononauts may one day drive their own Wellsian-type time machines, or glide through time portals that are controlled by technology rather than happening as unpredictably random anomalies of the natural universe. How else might we travel in and through time?

There is some neurological and psychological evidence that holds out the possibility that the *mind* — the real person, the sentient self, the soul or spirit — is independent of the physical brain. If this is correct, and if there exists within each one of us what some researchers refer to as an astral body, then, being non-physical, it could, theoretically, be independent of time. It could glide through the doors of time as easily as traditional Dickensian spectres glide through walls of stone. Dunne's dream evidence seems to recognize this possibility, and the data provided by genuine psychics and spiritualist mediums over many years points in the same direction.

One of our first-hand personal experiences of an astral time travel mystery goes back to 1997 when our immensely successful *Fortean TV* series was at its height on UK Channel 4. That show dealt with all kinds of mysteries, and there were many fascinating spinoffs from the series. One such spinoff led to our being taken to meet a mysterious and secretive inventor who was working in Bath, England. He had developed what he sincerely believed was a time machine. It was some time ago, and human memory is not infallible, but as far as we can recall, he reported that he had used the machine successfully for astral projection into the

Co-author Lionel with the inventor of a time machine.

Part of the time machine.

past, with quite traumatic results for at least one of his volunteers. He promised us a follow-up visit, when he would allow co-author Lionel to try the machine. Sadly, before that follow-up visit could be arranged, the old inventor died. Searching our records high and low for days — because his work on that machine would be so relevant to this chapter — we still cannot find his full name, although we *think* his Christian name was Gordon and his surname Pringele. But we did turn up this picture of him with co-author Lionel, together with a view of his intriguing machine.

There is another type of time travel, however, which is more accurately referred to as time observation. The prime example of this is the work attributed to Father Pellegrino Maria Ernetti (1925–1994), an Italian Benedictine monk who claimed to have invented and used a device called a chronovisor back in the 1950s. It is said that Ernetti was assisted in the work by a number of scientists, including Wernher von Braun and Enrico Fermi. If these accounts can be substantiated, then few scientists can have had better assistants.

Wernher von Braun was born on March 23, 1912, and died on June 16, 1977. He lived during the right period to have been involved with Father Ernetti and the controversial chronovisor. Von Braun's main interest and area of expertise was rocket science, and after the Second World War he worked on the American ICBM program, later becoming a director of NASA. His father, Magnus, served as Minister of Agriculture in the ill-fated Weimar Republic in Germany, and his mother, Emmy von Quistorp, was able to trace her royal ancestors back to Robert III of Scotland, Phillip III of France, Edward III of England, and Valdemar I of Denmark. With a family tree like that, it is not unreasonable to suggest that Wernher von Braun might have been interested in using a chronovisor to see what his royal ancestors had been doing in bygone centuries.

Enrico Fermi was born on September 29, 1901, and lived until November 28, 1954. His dates also put him safely into the period when he could have worked on the chronovisor with Ernetti during the early 1950s.

In assessing the reliability of the claims for the chronovisor, it has to be remembered that Ernetti was no mean scientist in his own right. He had a degree in quantum and subatomic physics, although his main interest was prepolyphonic music, which had been composed during the

Brilliant scientist Enrico Fermi.

three millennia from 2000 B.C. until medieval times. What might he have known about the strangely carved musical codes in Roslyn Chapel near Edinburgh, packed with its curious Templar riddles? Ernetti was also famous as an exorcist, and in great demand for his services in that area.

Almost inevitably, where there is a major mystery there will be arguments, controversies, and varying opinions. Some researchers into the mysteries of Ernetti's chronovisor maintain that he did not give true and accurate accounts of what it could do and what he could see in it. Others allege that on his deathbed, he confessed that some of his major claims about it were not genuine. Yet Father Brune — who knew him well and has written a great deal about him — regards Ernetti as a man of brilliant intelligence and high integrity. Brune suspects that it is the so-called deathbed denial that is fraudulent. He also raises the intriguing possibility that — like the strange mysteries surrounding Father Bérenger Saunière and the enigmatic treasure of Rennes-le-Château (which the authors have been investigating since 1975) — the Vatican may have Father Ernetti's chronovisor secretly hidden away somewhere. The chronovisor is described as being a large cabinet containing a cathode ray tube. It can be focused by using keys, buttons, knurled tuning wheels, levers, and other sophisticated control gear, and can be tuned to particular historical dates. It also has the power to track individuals who lived in the past and whom the user particularly wants to observe. Ernetti's basic theory was that all activities and events produce electromagnetic radiation, and he claimed that his chronovisor could detect it, decode it, and reproduce the sights and sounds that had created it in the past. He also maintained that he could

Musical codes in mysterious Roslyn Chapel in Scotland.

detect sound wave traces from the past as well as images produced by electromagnetic radiation.

A device like Ernetti's chronovisor — if it ever really existed — lends itself to the deepest, darkest conspiracy theories, and perhaps paranoia. Is it hidden in the Vatican? Is it in the hands of what conspiracy theorists regard as the secretive *real* rulers of the world? Or was Ernetti helped by extraterrestrials as well as brilliant human physicists when he put the chronovisor together? If it still exists, is it on Earth? The Ernetti mystery is a substantial one that cannot lightly be dismissed.

Time travel may also be possible as a psychic phenomenon — real astral travellers may be able to do far more than simply observe the past. Could astral travel actually take the disembodied voyager through time as well as through space? Is this another possible answer to the recurring problem of "If time travel exists, why haven't we met any time travellers?" An invisible astral time traveller might not be detectable to those inhabiting the past that he was visiting — or perhaps only a few unusually gifted and sensitive people could detect him. Some of the ghosts, spectres, or apparitions reported from Victorian séances might possibly have been time travellers from another age. Those that wore the garb of earlier centuries could have travelled *forward* astrally from their own real time to a

Bridge over the Tiber, close to the Vatican, where the mysterious machine may lie hidden.

Victorian era that was only one of an infinite number of future probability tracks as far as the travellers from pre-Victorian times were concerned. There could as yet have been no concrete and inflexible future from their point of view. The question becomes statistical again. If enough astral time travellers ventured out — knowingly or unknowingly — then some of them may well have appeared at nineteenth-century séances on a probability track that turned into an experiential reality, and ultimately became part of the solidified history of our twenty-first century.

What if an astral time traveller went back into the past? If a perceptive psychic witness observed a strange, disembodied entity during a séance tonight — an entity that was dressed in a fashion unfamiliar to our twenty-first-century eyes — how might that encounter be reported? Would the visitor be recognized as a time traveller or a guest from another dimension, or would it be recorded as a phantom?

The remarkable case of Mrs. Emma Turner of Lowestoft in East Anglia, United Kingdom, may shed some light on the question. We were conducting a series of lectures on the paranormal in Lowestoft when Mr. A.M. Turner told us the fascinating story of his great-grandmother. As we recall the episode, she and her future husband had apparently met in the United States while he was visiting there, and had become engaged. She came to the U.K. to go house hunting with her fiancé, and as they approached one particular country house set in its own spacious grounds, she told him that she knew this place well — although she had never been to the U.K. before. Because of several strange dreams that she had had of a beautiful old British country house, she was able to tell her future husband in clear detail what they would see as their carriage took them ever closer to the house itself — and she was correct. When they reached the front door and rang the bell, the butler froze and stared in amazement at the future Mrs. Turner. Regaining his self-control at last, he apologized profoundly and said, "I do beg your pardon, madam, *but you are the lady who haunts this house!*" The inside of the building was just as familiar as the approach to it had been. This version of the story ends well. Mrs. Turner and her husband lived long, happy lives in the house together, and raised a family there — one of whom grew up to be the grandfather of our friend A.M. Turner, who told us the story. Was it some strange precognitive incident that led Mrs. Turner to visit the house astrally long before she and her husband bought it? Did she have Dunne-like dreams of future happiness in that place, and did those precognitive dreams stimulate her astral body to go and have a look at

it while she slept? Like many of the best mysteries, however, there is more than one version of the events. An article in a 1910 issue of *Lowestoft College Magazine* gives the location of the house as Scotland, and makes no reference to Emma's being an American nor to her going house hunting prior to her marriage. However, the lady who is central to the account is not named in this article, and there could well have been *two* similar dream adventures. Emma Turner lived from 1840 until 1917 and lies buried in St. Margaret's Church cemetery in Lowestoft, where her gravestone can still be seen.

If astral time travel is one of the many weird facts that make up the enigmatic data of this strange

Roger Bacon, monk and scientist.

universe, does it take other forms? Is it possible for an astral time traveller to get caught or trapped in the past? Could that be one possible explanation for those geniuses who seem to be born ahead of their time? Does astral time travel explain Roger Bacon, Leonardo da Vinci, Isaac Newton, Nikola Tesla, and even Albert Einstein?

Judged by his ideas, which were centuries ahead of his time, Roger Bacon could very well have had a twenty-first-century mind trapped in a thirteenth-century body. Born in 1214, Bacon was a pioneering scientist with a deep belief in the power of reason. In his famous book *De Mirabili Potestate Artis et Naturae* ("*On the Marvellous Powers of Art and Nature*"), Bacon argued that the human mind was so powerful that it could do the seemingly impossible — such as designing and constructing a machine that would enable a human being to fly. He then outlined two broad plans for achieving such flight. The first was an ornithopter, a machine that, if provided with large enough wings and sufficient power to flap them effectively, would fly. The second was a lighter-than-air idea. Bacon argued

Malmesbury Abbey, where a medieval monk once flew.

that air was a kind of fluid in which a globe filled with air that was "thinner" than normal air would float, just as a ship floats on water. He went on to suggest ways of thinning the air in the globe so that it would become buoyant. Bacon then makes a mysterious statement about an early aviator: "There is an instrument with which it is possible to fly. I have not seen it, nor do I know any man who has seen it — but I know the name of the wise man who created it."

Bacon was possibly referring to the fearless young monk Eilmer — sometimes called Oliver —

Eilmer: the flying monk of Malmesbury with his wings.

from Malmesbury Abbey in Wiltshire, United Kingdom. Born *circa* 980, Eilmer launched himself from the top of the Abbey tower with wings similar to those described in the Greek legend of Daedalus and Icarus. Eilmer seems to have been airborne for about twenty seconds, but broke both legs when he landed. It is possible that Eilmer had heard stories from returning Crusaders of the flight accomplished by Moorish inventor Abbas Ibn Firnas from Cordoba, Spain, in 875. Abbas had successfully glided several hundred feet with wings made roughly on the Daedalus pattern. William of Malmesbury includes an account of Eilmer's aerial adventure in his book *Gesta Regum Anglorum* (*"Deeds of the English Kings"*).

Bacon is best known for inventing gunpowder, and he wisely hid the formula for it in a Latin cryptogram. Scholar and scientist that he was, Bacon described the results dramatically: "The explosion I have made roared louder than thunder and its flash was more brilliant than lightning."

In addition to his advanced chemical experiments, Bacon wrote knowledgably of a magnifying glass to help with reading, and he also predicted aircraft, cars, and powered ocean-going ships. It is not beyond

Leonardo da Vinci: was he a time traveller?

the bounds of possibility that this remarkable thirteenth-century monk-scientist was hosting a time traveller's mind. The other factor that put him so far ahead of his time was his advocacy of the modern, scientific attitude: "Never take anything for granted — observe with your own eyes and test every new theory with your own hands."

There may well be greater mysteries attached to the brilliant Leonardo da Vinci than the intriguing and controversial religious code with which Dan Brown has associated him.

There are several unresolved questions about Leonardo's birth on April 15, 1452. One school of thought suggests that he was the illegitimate son of a wealthy Florentine and a beautiful peasant girl from Anchiano near the little town of Vinci in Tuscany. Another theory gives Vinci as his mother's name, rather than the name of the town. Academic historians offer his full name as Leonardo di ser Piero da Vinci, meaning Leonardo son of Piero from Vinci. He grew up to be an outstanding polymath, a true Renaissance man. He was a brilliant painter, sculptor, engineer, and inventor, as well as an architect, scientist, mathematician, musician, and anatomist. It is his inventions, however, that are of most interest in our present context. He designed a parachute, a helicopter, a military tank, and a diving suit that was clearly the forerunner of modern scuba kit. Could he, too, have been a twenty-first-century mind trapped in a fifteenth-century body? Da Vinci died on May 2, 1519 — centuries before his inventions came into everyday use.

Isaac Newton (1642–1727) is generally regarded as one of the finest intellects the human race has yet produced. Primarily a mathematician

Leonardo's battle tank — top and base.

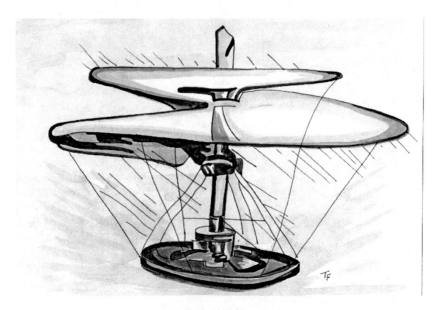

Leonardo's helicopter.

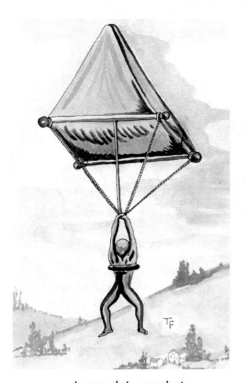

Leonardo's parachute.

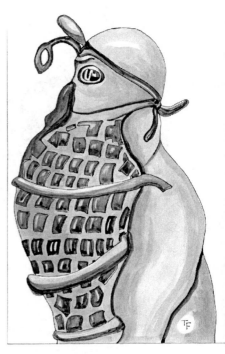

Leonardo's scuba gear.

and physicist, Newton was also a deeply religious man, very interested in Judeo-Christian prophecies, which he thought were essential to any proper understanding of God and the universe. As detailed in Chapter 2, Newton had a strange start to life, which has certain broad parallels to Leonardo's birth circumstances. Did both Newton's and Leonardo's great minds come from a time ahead of their own?

If Bacon, Leonardo, and Newton pose interesting questions about the possibility of astral time travel, they are overshadowed by the mysteries associated with Nikola Tesla.

If the legends surrounding Tesla's birth are true, his arrival on Earth was stranger than Leonardo's or Newton's. He was born exactly at midnight during a wildly fierce electrical storm. The date was July 10, 1856. His father was Milutin Tesla, a priest in the Serbian Orthodox Church. His mother, Duka Mandic, was the daughter of another Orthodox priest. Whatever genes Nikola inherited from her would have helped his future career as an outstanding inventor: Duka was expert at designing and making domestic equipment, and she could memorize lengthy epic poems. Having inherited her talent, Nikola could memorize entire books, and he also developed a form of visual thinking that enabled him to construct an invention realistically inside his mind before making a solid, physical prototype of it.

His career as an electrical inventor verges on the miraculous. His creations — either invented from scratch or improved and developed by Nikola Tesla — included many advanced devices that depended on rotating magnetic fields; the induction motor; rotary transformers; the Tesla coil; the Tesla oscillator; alternating current long-distance transmission systems; radio frequency oscilla-

Nikola Tesla, the great electrical scientist.

tors; forms of commutators; bladeless turbines; and voltage multiplication circuitry. In addition to all of these, Tesla produced artificial lightning with millions of volts and sparks 135 feet long. He recorded what he believed were signals from extraterrestrials, and he designed a deadly death ray (which he called a peace ray because it would totally defeat the strongest army). However Tesla is judged, he could certainly be described metaphorically as a man ahead of his time. Could he have been a twenty-first-century mind trapped inside a nineteenth-century body?

The name of Albert Einstein is probably the first that comes to mind when we look for a person who can readily be associated with the concept of human genius. Einstein, a worthy Nobel Prize winner for physics in 1921, was an independent thinker, and that intellectual autonomy was allied with massive determination and an indomitable will to succeed. Most famous for his work on relativity, Einstein was often described as a man ahead of his time. He is undoubtedly among those outstandingly able thinkers, like Bacon, da Vinci, Newton, and Tesla, who could have come from an advanced scientific future.

If the previous examples present interesting conjectures about time-transcending visitors, the story of John Titor, who posted on the Internet, claiming to be a soldier from 2036, is an even more remarkable one. He described his time machine as something along the lines of a Tipler cylinder, operating with miniature black holes capable of distorting the gravity in their immediate areas and, in doing so, allowing a chrononaut to move through time. The Titor machine would therefore seem to combine the properties of a Wellsian device that the pilot can control with a time portal opener.

Titor's statements about the future are not encouraging: civil war in the United States; a cataclysmic nuclear holocaust in 2015; gradual recovery afterwards so that by 2036 science and civilization were able to send him back to our time to warn us. If John Titor was a genuine time traveller, his message reinforces the theory of infinitely branching timelines and probability tracks. A change for the better in human behaviour *now* might steer us away from the 2015 disaster that he says was part of the history of his experiential world. John Titor could be only a clever hoaxer — but he could just be what he claims. Only time may tell — and it won't if we move to a more benign experiential actuality. There lies the quaint and curious paradox of branching timelines and probability tracks. Someone like Titor, for example, returns to an earlier time and warns his predecessors that horrendous consequences lie ahead of them. They listen to his warnings and take effective evasive action: disaster doesn't strike. Because the world lines have diverged there would seem to be no way of proving or disproving the supposed time traveller's message — unless a means could be found of travelling *between* the diverging world lines.

In Chapter 5 we examine the great issues that are raised by the philosophy of time.

THE PHILOSOPHY OF TIME

GETTING TO GRIPS with the philosophy of time necessitates looking into general philosophy first. Professional philosophers tend to regard the definition of philosophy as one of the major hurdles that has to be scrambled over in pursuing their vocation. It was Socrates who said that the unexamined life was not worth living, and philosophy may be defined as one of the major intellectual instruments enabling us to carry out that essential examination.

The *Oxford Dictionary of Philosophy* makes the interesting comment that most definitions of philosophy are controversial, and other academic experts suggest that defining philosophy is notoriously difficult, or that no satisfactory definition yet exists. The authors, however, would define philosophy as the most effective weapon yet devised for use in the human armoury during the interminable struggle to understand our own meaning and purpose in the light of the incomprehensible phenomena both within us and around us. Philosophy is the quest for the special kind of knowledge that encapsulates and attempts to give meaning to everything of which the human mind is aware.

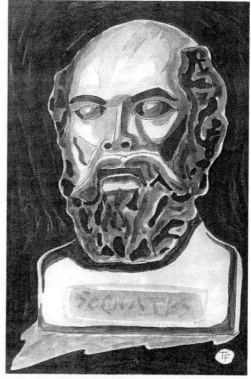

Bust of Socrates, one of the world's greatest philosophers.

Philosophers take pride in the logic and reason with which they pursue their inquiries. If philosophy itself cannot easily be defined, its methodology is less elusive: philosophy employs rational, logical arguments and is by nature both systematic and critical in its approach.

If that is its *method*, what is philosophy's intrinsic *nature*? It differs from science because it cannot use empirical methods. There are no philosophical laboratories. Philosophy cannot be shaken up with dilute acids in test tubes. It cannot conduct electric currents. Its genes cannot be engineered. Just as surely as philosophy differs from observational and empirical science, it also differs from religion: it has no commitment to faith, nor does it believe in any divine revelation. Philosophy steers its stately craft by the guiding star of intellect alone.

It is also germane to say that philosophy is one stage removed from what may be described as *first-order* studies. Historians study the past that has shaped our present. Geographers study the planet we live on. Astronomers study the stars. Mathematicians revel in the intriguing relationships of numbers, spaces, and shapes. Linguists examine oral and literary expression. These first-order disciplines necessitate examining their own particular subject matter directly, forming theories about it, initiating relevant new concepts, making observations, and working with presuppositions allied to the subject matter being studied. Philosophy doesn't do that. Philosophy is a *second-order* study. It can be starkly abbreviated to the idea that it is essentially thinking about thinking. Whereas the first-order disciplines require thought about their subject matter, philosophy requires thought about the nature of thought.

For example, how does a mathematician solve this intriguing little problem? It needs first-order mathematical thinking. Imagine a kindly visitor to an aquarium. He brings a tasty meal consisting of nineteen fish to be shared among walruses, dolphins, and penguins. He and the aquarium keeper decide that every walrus deserves 1.5 fish, every dolphin shall have 1.25 fish, and every penguin will be given 1.2 fish. As the lucky recipients are in different enclosures, the whole fish need to be divided up among the enclosures. The mathematician, tackling the puzzle, has to work out how many of each species are in the aquarium and how many fish go to each enclosure. There must be the right number of fish placed in each enclosure to provide the appointed fractional shares

THE PHILOSOPHY OF TIME

for all the lucky denizens of that enclosure. The mathematician begins by working out that there must be 3, 6, 9, or 12 whole fish for the walruses, so that the total fish can be divided exactly by 1.5. To deal with the 1.25 fish ration, there must be 5, 10, or 15 whole fish given to the dolphins. The penguins must have 6, 12, or 18 fish. Having worked that out, the mathematician then goes through the three separate series of multiples that he has constructed and takes one from each sequence so that they add to 19. As 3 + 10 + 6 gives the required total of 19, there must be 2 walruses accounting for 3 fish, 8 dolphins eating 10 fish, and 5 penguins enjoying 6 fish.

A philosopher, observing the mathematician, is interested in the *way* that he thinks, not in the answer to the puzzle of the nineteen fish. Philosophers examine the concepts and theories that mathematicians — and other first-order intellectuals — create. Philosophy can also be identified as a reflection about *styles* of thinking.

In examining philosophy as a second-order discipline, it is important to recognize that the boundaries between first- and second-order thinking can sometimes be hazy and indeterminate. Philosophers of science may come up with ideas about scientific thinking as a process, which may stimulate a first-order scientist to approach an empirical problem in a different way and solve it by using a different process. Conversely, hands-on scientists may make discoveries that contribute to the philosopher's understanding of the scientific method, and go at least part of the way towards resolving a difficult paradox or an apparent contradiction in the philosophy of science.

It is also intrinsic to philosophy that its etymology is not helpful — at least on the surface. Breaking the word *philosophy* down to its oldest roots gives the meaning "a love of wisdom." But do all philosophers have real affection for deep and purposeful wisdom? The best ones do, of course, but just as the sophists plagued ancient Greek thought with their superficial trivia, there are contemporary philosophers who amuse themselves by merely playing tricks with words or juggling apparent paradoxes and contradictions. The vitally important task of the true philosopher is to seek for meaning, and, having found it, to share it with the rest of us.

There is also an intrinsic element of the highly desirable critical and questioning Fortean attitude in philosophy: philosophers take nothing on trust. The more obvious something seems, the more avidly philosophers will question it — and that is excellent.

Another approach to an understanding of the intrinsic nature of philosophy is to ask what it is *not*. Some academic philosophers might argue that their discipline is not about an attitude to life — yet hedonism, stoicism, and skepticism all seem to have a profound influence on their followers' lifestyles.

Moving on from the intrinsic character of philosophy, further insights into its real nature and function can, perhaps, be gained by examining its subject matter. Philosophy looks at fundamental things. It looks in the widest and most general sense at the entire universe and the role of humanity within it. Philosophers survey causes and effects. If a question is big enough, philosophers will look at it. Philosophy digs fearlessly at the deepest of all intellectual levels: it has truly been described as "the study of ultimate reality."

The attempt to understand philosophy as a whole can be enriched and enhanced by looking at its branches and subdivisions. It incorporates metaphysics — the study of things like time and infinity that go beyond the simple here and now of everyday life. Philosophy also includes epistemology (sometimes called the theory of knowledge). Epistemology asks what knowledge is, what types of knowledge exist, and how we acquire and process what we think of as knowledge. Epistemology challenges the philosopher with three fundamental questions: What do we really know? What is this thing that we call knowledge? How do we obtain it? One of the subtle distinctions between the different types of knowledge is what we might call *knowing-that* (for example, knowing that $7 + 3 = 10$) and *knowing-how* (for example, knowing how to add 7 units to 3 units and end up with 10 units). A chemistry teacher may know that adding sodium metal to chlorine gas will produce sodium chloride, or common table salt ($Na + Cl = NaCl$). But both sodium and chlorine in their uncombined states are very dangerous to work with, so the teacher has to make sure that the students know how to take all the necessary safety precautions before performing experiments of that kind. Essentially, the two different types of knowledge can be regarded as being parallel to theory and practice. Epistemology is associated with theory.

Epistemology is also concerned with ideas of *belief* — but with specific forms of belief. If a boxing supporter believes that an up-and-coming young fighter will become a world champion one day, he is making a mental prediction about the skill and ability of that boxer compared to other fighters. If a builder believes that a ladder and scaffolding are safe and strong enough to take his weight, but they break and he falls, his belief was not correct. He did not have *knowledge* that they would hold him — he only *believed* that they would. In epistemology, something has to be true in order to count as knowledge. There are epistemologists who would contend that knowledge is true belief that has been justified.

The problems associated with philosophy, and its second-order nature as thinking about thinking, are refined even further when metaphilosophy comes into the spotlight. Metaphilosophy can be described as "the philosophy of philosophy." The primary focus of metaphilosophy is what we have been describing in this chapter: it asks what philosophy really is.

Having looked briefly at the nature of philosophy and some of its activities and purposes, we are in a better position to focus on the philosophy of time itself.

A central point in the philosophy of time — argued with due deference to the rational logic that is central to all good philosophy — is whether or not time can be considered to have had a beginning. A well-known philosophical objection to time's having had a beginning is that in order to exist a thing needs to have had a beginning, and it is a truism that everything that begins must begin in time. If that is correct, it is argued that time itself cannot have had a beginning. This premise seems to rely on the idea that a *beginning* means that there was an earlier time at which the thing under consideration did *not* exist, and a later time at which it *did* exist. If this argument is pursued, it seems in some minds to lead to the conclusion that it is impossible for time itself to begin because that would imply that there was a time earlier than the beginning of time! Opponents of this argument suggest that it all depends upon what we mean by the concept of *beginning*. It can be argued that when we apply the concept of *beginning* to time it has a different sense from that which is understood when the concept of *starting* is applied to things or states of being. It has been argued that when the idea of *beginning* is applied to time, it suggests that there is a measurable interval of time right at the very start — so that every other moment in time will be at some later time than this original one.

Another way of looking at the philosophy of time — and the beginning of time — is to suggest that unless all the time intervals involved in the argument are of the same length, then the first second will have elapsed before the first minute elapses, and that will have gone before the first hour has passed. This can apply to the first day, the first week, the first month, the first year, the first decade … and so on until the largest conceivable units of time have all elapsed. The problem with this sequence of units of time of different lengths is that the second is not the shortest interval — we have microseconds, nanoseconds, and so on into ever smaller divisions of duration, until we run into the buffer of time, the instanton … if it exists!

If time is incapable of beginning, if there is no point in the history of the cosmos at which it is impossible to describe an earlier point, then it may be argued that time is infinite — insofar as it has no start and no finish. If that is true, then, perhaps both time and space are boundless — and the philosopher is thrown back into the company of the most daring cosmologists who talk of four-dimensional space-time continuums and string theory.

But the most important aspect of the philosophy of time is the way that it affects us as thinking, caring, sentient beings. Just as we noted that hedonism, skepticism, and stoicism have very pronounced effects upon the thoughts, actions, and lifestyles of those philosophers who hold those theories to be true, so our philosophy of time is likely to affect the way that we think, speak, and act during our existence in time.

For adherents of many major world religions (including Christianity) the afterlife is seen as eternal and infinite, not only quantitatively but also qualitatively. They believe that the life that follows our earthly existence is not only everlasting but abundant as well: it is not only endless but also richer, fuller, and more worthwhile and fulfilling than any experience that the happiest and most exuberant human being has ever known on Earth. For believers, the life of heaven, or paradise, extends joyously in *every* direction — not just in terms of its infinite length. There is nothing in objective science — nor in rational logic and philosophy — that can definitively refute that religious view. In fact, there is powerful support for it to be found in both.

If our chosen philosophy of time is one that accepts the reality of eternity, then we will think, speak, and live as if we expect to participate in a gloriously happy, unending future state.

Suppose instead that our philosophy of time is one that sees time as having both a beginning in the past and an end in the future: how is that philosophy of limited, circumscribed time likely to affect our thoughts, words, and deeds in the here and now?

No matter how far ahead we may envisage the future ending as taking place, there are some adherents of this finite time philosophy who would feel that even if the end of time is delayed for billions of years, it still gives a sense of emptiness and futility to life now. If the greatest architecture and the finest music, art, and literature produced by human thought and endeavour will all end one day as though they had never been, what's the point of creating them at all? If time ends, and death and oblivion are the only future for the universe and its inhabitants, what's the point of being part of it? The answer must surely be that there are things which transcend time and which have an intrinsic value out of all proportion to their duration. These include sentient thoughts, feelings, and relationships. We would say that to have loved and to have been loved — and to know it unquestionably at the deepest possible level — possesses that kind of transcendence. Lovers know instinctively that what they feel for each other is infinitely precious and independent of duration. In an ideal universe, it lasts forever. In the hazardous universe we live in, the fragility of human life can tragically interrupt it. Yet lovers who enjoy relationships in which the life of their beloved is far more important to them than their own have touched something so exquisitely powerful that its effect is permanent. Its mark on them is indelible. By experiencing that intensity of love, they have — in a very real sense — escaped from the limitations of time into eternity. We have already argued that eternity is qualitative as well as quantitative. It can, therefore, be argued further that to have experienced its qualitative aspect is to qualify as a participant in its quantitative aspect. In our philosophy of time, the quality and quantity of abundant and eternal life are inseparable.

In addition to love, the sublime thoughts of philosophers, mathematicians, theologians, scientists, poets, musicians, artists, and architects may also transcend the framework of terminable time.

If our philosophy of time is linear, the inevitable time arrow pursuing its remorseless unidirectional course from the past via the present to the future, then our whole mode of thought and action may well be permeated by this concept of linearity. The thinker who is governed by linear time will tend to think of critical paths of cause and effect, and to have an attitude of "What I did yesterday has affected today, so what I do today will affect tomorrow." Such thinkers are likely to be both pragmatic and dynamic. They will look at their immediate, intermediate, and distant goals and objectives and plan such actions as will, in their opinions, most effectively achieve those aims.

If our philosophy of time is circular or cyclic, our mode of thought is likely to be more religious, mystical, and spiritual. Philosophers who embrace theories of cyclic time are likely to see the cosmos — and their own lives as part of it — as rhythmic and following one another in an endlessly recurring sequence. They see the rolling seasons — spring, summer, autumn, and winter — as smoothly repetitive, and as symbolic of life itself. In their minds, the changes that come and go will eventually bring back a golden age of peace and prosperity. Unlike the dynamic and purposeful linear time philosophers, the cyclic time thinkers are relaxed and accepting. They acquiesce to what they see as the universe's revolving nature: *que sera sera* ("whatever will be will be") — and, furthermore, it will be again as it was in the past. Does it go from the Big Bang to the ultimate expansion limits then shrink back to an infinitely dense singularity, followed by another Big Bang ... and again ... and again ... ad infinitum? Cyclic time theorists have a tendency to see the cosmos throbbing regularly, like an unimaginably large heart. Cyclic theories of time are likely to lead to acceptance of déjà vu phenomena and reincarnation.

Another theory of time considers it to be infinitely branching. At each moment of time there are innumerable possibilities. Following track one leads to a potential future different from that which might have come to pass had track two been selected instead. This philosophy of time suggests that there are infinite probability tracks ahead of us at any given moment.

The linear time arrow theorist feels that today's choice of action gives him some control over tomorrow. He works on the assumption that his plans and actions affect his future.

The philosopher who goes with the theory of perpetually bifurcating time, however, sees his decisions and actions in the present not as controlling or influencing the future, but as opening or closing an infinite number of probability portals through which he *could* pass. Current actions open and close the entrances to different probability tracks that will become "real" and experiential tomorrow, depending upon today's choices.

Philosophers who hold the bifurcating time theory will also be keenly interested in the mechanisms that they think *might* be able to connect the variant probability tracks. For example, a candidate may have turned down the offer of a university education or a promising job and later regretted that decision. He is now on an unwelcome reality track with low income and few prospects. The missed opportunity would have opened the probability portal to much greater satisfaction on another track. The individual thinks things through and decides to go back to the point at which the wrong decision was made. He clearly recognizes the impossibility of backtracking through time along the current experiential reality track to the moment at which the opportunity was offered. What is possible, however, is to build a "bridge" further along the time-stream and to reapply for the university place or the trainee manager's position. If the reapplication succeeds, then the applicant is heading for a point where the tracks will converge at some future time — and he will be more or less on the same experiential track that they would have been on had the better decision been made earlier. For such a bifurcating-time philosopher, a lifestyle of looking for — and taking — second chances and postponed opportunities is characteristic. For a person with this philosophy there is no such wistful thought as "If only I'd ..." or "I wish I'd done X instead of Y, *then* ..." Bifurcating-time philosophers believe that they should never give up and they continually look for ways of entering better probability tracks than the one they are currently on.

Finally, we need to consider the integrated-time philosopher: one who goes along with string theory, or with the earlier Einsteinian view of a four-dimensional space-time continuum. This particular time philosophy tends to colour its holder's thoughts, words, and actions with the feeling that the cosmos is a unity, a totality, of which the philosopher is an essential and meaningful part. Space, time, and energy are seen as the foundation, the vital skeleton that supports the cosmos and all the sentient life forms in its various biospheres. The integrated-time philosopher is happiest as a team player, a committee member, or a collegiate worker who acknowledges the importance of the other team members; she derives a sense of worthiness and personal value from knowing her place in the team. For the integrated-time philosopher, there is no difference in value between time past and time to come. The present moment is the experiential viewing port through which time can be glimpsed and revealed as an integral part of the cosmic structure.

In considering the philosophical arena in general, and the philosophy of time in particular, the authors tend to diverge from the view that philosophy is not concerned with the way that a particular theory affects the holder's thoughts, words, and lifestyle. In our opinion, the clear and unequivocal recognition that a favoured philosophy can and does affect the way we live is of some real importance. Because we live in time, and because time dictates what has to be done — or *ought* to be done — during certain intervals in our existence, we need to decide on our philosophy of time and then allow for the influence that that particular philosophy of time may be exerting on us.

Whether we favour the theory of terminal time, everlasting time, cyclic time, bifurcated time, or integrated time is less important than our conscious *awareness* that different philosophies of time exist, and that one of them may be influencing our lifestyle. It is prudent, therefore, to think carefully about which — if any — of these time theories we prefer. The five different philosophies of time cited in this chapter are by no means the only ones in existence, and it may well be that the creatively thoughtful, analytical reader can suggest others that are equally rational and equally probable. In looking into such new theories, however, it is equally important to recognize that — as with those described and analyzed in this chapter — the philosophy of time that each of us holds is capable of influencing our thinking and our behaviour.

In the next chapter we move on from the philosophy of time to the theology of time and eternity.

Chapter 6

THE THEOLOGY OF TIME AND ETERNITY

"Thy throne is established of old: thou art from everlasting."
— Psalm 93, verse 2.

THE THEOLOGY of most of the major world religions regards God as predating time by a factor of infinity. Distinguished theologians of all the major monotheistic faiths tend to think of God as the creator of time, the sustainer of time — its director and controller. The biblical Book of Daniel refers to God as the "Ancient of Days." In Aramaic it is rendered *atiq yomin*, in Greek it is *palaios hemeron*, and in Latin *antiquus dierum*. It carries the sense of an ultimate god who predates time, transcends it totally, or is understood to be so infinitely powerful that he is impervious to it.

In Ecclesiastes 3:1–6, there is also an implied sense of particular times and seasons being allocated by God for particular events:

To everything there is a season and a time to every purpose under the heaven: a time to be born, and a time to die; a time to plant, and a time to pluck up that which is planted; a time to kill, and a time to heal; a time to break down, and a time to build up; a time to weep and a time to laugh; a time to mourn and a time to dance; a time to cast away stones, and a time to gather stones together; a time to embrace, and a time to refrain from embracing; a time to get, and a time to lose; a time to keep and a time to cast away …

This interesting piece of what biblical scholars categorize as Old Testament wisdom literature raises the contentious issue of prophecy and predestination. Academic theologians ask whether the author of Ecclesiastes intended his words to mean simply that (as part of the divine wisdom) various events in life were planned, regulated, and apportioned like the seasons and phases of the moon, or whether he was indicating that the precise times of birth and death, killing and healing, gaining and losing were all predestined for each individual person.

The doctrine of predestination automatically raises profound moral and ethical difficulties: awesome philosophical and theological conundrums are an inescapable part of predestination. If time has such a powerful consistency that it can hold events as enduringly as carved marble can hold images and inscriptions, and if those events that are as yet hidden from us in our future bear down upon us inevitably, then morality and ethics, vice, virtue, and personal responsibility have no meaning. Jack the Ripper, Bluebeard, pirate Captain Teach, Hitler, and Mussolini are neither more nor less virtuous than St. Francis of Assisi, Florence Nightingale, Pope John Paul II, or Mother Teresa. If time is a single, arrow-line progression along which each of us plays a predestined role from which we cannot deviate, then our existence has no moral meaning. We are merely sentient raindrops falling from clouds in accordance with the law of gravity and the local air pressure and wind direction. We are as devoid of choice — and of ethics — as boulders crashing down the slopes of Snowdon.

However, if time is bifurcated in some way, splitting into one probability track after another — giving us millions of moment-by-moment choices and opportunities to do good or to do evil — then ethics and morality certainly exist. Without individual freedom of choice and individual responsibility for our behaviour, life would seem pointless. If we are merely time's helpless automatons, tragically troubled by an awareness of our own existence, then some philosophers and theologians would conclude that we might as well not be here at all. But whatever else the wisdom literature means, in our opinions it does not mean predestination. God invites: he does not compel. The higher theological concept of God as master and controller of time does not make him the author of inevitable events. It is far closer to the truth to describe God as the ultimate hero, who has dared to give us the awesomely dangerous gift of free will.

The idea of such a loving, caring, freedom-giving God, the ultimate master of time, is in sharp contrast to the ancient Greek idea of Chronos, who was regarded as the personification of time. In pre-Socratic thought, he emerged from Chaos (the random and disorganized state of everything that preceded creation, according to this ancient mythology). Greco-Roman art shows him as the man who turns the wheel that holds the Zodiac. Chronos is also known as Aeon, which was thought of as *more* than transient time. Aeon seems to have personified the concept of the ages, eternal or everlasting time. There are subtle elements of Aeon that are stronger than those of Chronos. Aeon is not a relatively harmless elderly man with a long white beard — the popular image of Father Time.

Chronos's connection with the Zodiac, and the idea of the god who turns the wheel, led to his being associated with the planet which is now known as Saturn, but which the early Greeks named Chronos. The limitations of their astronomy saw Saturn as the outermost of the seven planets they knew of. To them, that planet also had the longest observable movement (approximately thirty years), and this may have contributed to their associating it with time. If — though it seems unlikely — any early Greek astronomer was aware of the ring system circling around Saturn, then it would have been an even more compelling reason for them to have associated that seventh planet with the god who turned the wheel of the Zodiac.

Another ancient Greek theological concept regarding time was that Chronos was sometimes regarded as the *keeper* of time, a watchman and guardian of it. Tangled inextricably with these time god and time watcher concepts in early Greek thought was the idea of Cronus, ruler of the Titans, who was seen as a Titan god of time.

Yet another fascinating fact here is that Chronos (alias Aeon) was associated in Greek mythology with Ananke, the goddess of inevitability. According to the Greek mythology of creation, she had somehow managed to emerge of her own volition — simultaneously with Aeon — right at the very beginning of time. Often thought of as a vast, snake-like form, but with arms that extended across the entire universe, Ananke was seen as inextricably entwined with Aeon . The two of them were in charge of the primeval "egg" whence came, under their direction, the land, the seas, and the sky. Ananke, the goddess of inevitability and her partner, the god of time, were thought of as ultimately controlling everything: not even the Greco-Roman gods could evade their powers.

Whether viewed from the standpoint of the priests, prophets, and law-givers whose thoughts contributed to the beginnings of the major monotheistic religions, or from that of the mythologizers who attempted to explain time in pictorial ways, the significance of time and its enigmas has presented inescapable problems to all speculative thinkers.

Yet another angle on the theology of time is the concept of time's apparent variability. Shakespeare comments on it as ambling, trotting, galloping, and standing still in Act III, Scene 2, of *As You Like It*. He is looking here at the subjectivity of time as it is experienced by different human minds in different life situations. Boredom drags time out to intolerable lengths, as does the torture chamber. A victim experiencing hot irons, the rack, or the brodequin will feel as if every second lasts a thousand years — whereas long, happy hours of pleasure and excitement seem like just a few seconds. But to the mind of an immortal and omnipotent God with absolute power over time, time itself becomes infinitely flexible. In the words of the old hymn "O God, Our Help in Ages Past" we find the lines: "A thousand ages in Thy sight are like an evening gone,/Short as the watch that ends the night before the rising sun."

I Corinthians 15:52, talks of the end of the world and the start of abundant, eternal life: "In a flash, in the twinkling of an eye ..."

Yet stepping out of time and into eternity raises the question of whether escaping from the sequential flow of time that we experience here on Earth (whether that subjective experience accords with the true scientific reality of time or not) will prevent us from carrying out any activities at all in eternity. What has to be considered is whether our mental culture of *past>present>future* is essential if actions of any kind are to take place. It is the unidirectional time arrow of *past>present>future* that makes cause and effect an integral part of our understanding of life on Earth. A carpenter wishing to make a table begins by selecting the lumber; next he cuts the shapes he wants from it; then its components are joined together; finally it is polished. By the time he starts to polish it, the first three stages have moved into the past. It would not have been possible to polish the unprepared lumber. Stages of the activity have to follow one another in an ordered sequence of construction. Common sense (based solely on our earthly experiences of time moving along in the *past>present>future* sequence) tells us that if there is no time — no medium through which activities can move in a necessary and logical order — activities as we know them in the here and now of everyday, terrestrial life cannot take place. So, is eternity to be simply an eternal *now*, not an unending series of exciting, adventurous pleasures, each greater and more wonderful than its predecessor?

There are theologians of different faiths who would express the idea of heaven or paradise as just that: a state of being in which the participants are aware of one another and conscious of shared happiness and fellowship with God and with loved ones, but nothing more — just infinite and unchanging joy, tranquility, and contentment. But could such a *motionless* state — even a perfect one — truly satisfy the questing spirit of adventure that stimulates so many of the bravest and best human hearts? Such a paradise would be totally free of risk, but is there something inside some of us that feels the need to take risks?

It may also be conjectured that as time — like everything else — is within God's absolute control, those who have stepped out of time via the doorway of physical death may then have access to every age of history and every state of being: past, present, future, and *possible*. Looked at in that way, the time-free afterlife may be thought of as the most luxurious tourist hotel imaginable — from which those who have died may venture wherever they want to go along every avenue of time, space, and probability. Time is now their servant, their vehicle, their magic carpet. It can take them anywhere and allow them to observe or experience whatever interests them. But would even that be satisfying to the truly inquisitive and adventurous souls? If their ultimate "tourist centre" is a static state of being from which their adventures are more like watching or acting in films, rather than having *real* adventures — in which it is possible to get killed or badly injured — is there any real satisfaction in them?

May it be conjectured theologically, therefore, that eternal and abundant life will need to have an element of *real* progression? If the experience of passing from this earthly life to a better, happier, and infinitely more wonderful existence is to be truly satisfying, then won't it be necessary for that transcendent existence to have within it an awareness of something akin to the *past>present>future* sequence that we are used to here?

We are minute parts of an infinitely large cosmos. With everything that we were here on Earth improved and enhanced out of all recognition, is it not possible that there will be amazing new things for us to do that are real — not mere heavenly "tourist attractions," but real jobs, real quests, real adventures that last forever? That's the kind of eternal and abundant life that would most appeal to us!

Theologians may be thought of as the most venturesome and intrepid of all thinkers: with human minds (even the most brilliant ones) they dare to attempt to study the person and purposes of God. They look first into God's very nature and existence and try to work out what *manner* of being God is. Theologians who work within the belief structures of the great monotheistic world religions regard God as omniscient, omnipotent, and omnipresent. It is this factor of God's omnipresence that impinges upon the theology of time. Omnipresence does not limit God's being to an infinity of multi-dimensional space — it proclaims God's presence throughout past, present, and future *time*, and on every probability track. Theologically, an omnipresent God must exist in *every* time and in *every* realm of possibility — whether those realms are actually inhabited by sentient, experiential beings or not.

Omnipresence also implies God's eternal existence: God was always present. God existed *before* time ran, and, if time is finite, God will continue to exist *after* time vanishes. Omnipresence implies neither ending nor beginning.

Similarly, it can be argued theologically that God's omniscience, God's total and absolute knowledge and wisdom, make him infinitely superior to time, the perfect master and controller of it. God holds time in all its aspects safely and securely inside the divine mind. Time is simply an integral part of the divine structure of our cosmos — and of any other cosmoses that we don't yet know about, which are expressions of God's power and delight in creation. With infinite speed and infinite precision, the divine mind in its complete omniscience worked out exactly what time was to be, how it was to work, and how it was to serve the divine purpose. Whether cyclic, linear, integrated, finite, or infinite, theologically, time is exactly as it should be, and is as much a part of the divine plan as the three known dimensions of space, physical matter, and energy or forces.

When considering time juxtaposed against the nature of God, theologians would next consider God's omnipotence. If omnipresence and omniscience place time firmly under God's control, omnipotence has an even stronger claim over it. Theologically, omnipotence is absolute power, and absolute power must include power over everything that exists — including time. If nothing is outside the power of an omnipotent God — and theologically nothing can be — then time must be subject to God's power.

But there are other significant theological arguments that have to be entertained within this debate concerning God and time. The first of these is the theological concern with the *purposes* of this omniscient, omnipotent, and omnipresent God. The three unlimited qualities that define God's knowledge, power, and location are God's *attributes* rather than God's *nature*. Christian theologians, basing their thoughts on the teachings of Christ, would say that God's nature is love personified. Some academic theologians (including co-author Lionel) would add that true love is the greatest good in the universe, and that it can exist only if there is freedom of choice.

Love cannot be compelled; it can only be invited. It cannot be forced; it can only be given. God, the author of love and the personification of love, designed ideal love to be a reactive and interactive dynamism, a two-way process that seeks above all to give happiness to the beloved, and in so doing — because of the very nature of love — the giver receives even greater happiness in return. Just as the basic laws of physics reveal that on a suitable surface, a dropped object possessing elasticity will attain a series of ever-decreasing heights as it rebounds with less energy, so real love does the *exact opposite*. Each time true and unselfish love is given, its nature is to return in greater magnitude and with more force. Kindness is designed to amplify kindness. Generosity is structured to enhance generosity.

But free will —the essential prerequisite of love — can, paradoxically, cause horrendous damage to this ideal love-building and love-amplifying process. Trinitarian theologians would suggest that within the triune personality of the godhead in which they believe, divine love is perfectly reciprocated: increasing and expanding eternally among the three persons of the Trinity. Outside the perfection of the Trinity, however, there are massive obstacles and impediments to the expanding, reciprocating dynamism of love. Because of free will, love is *not* universally reciprocated. An adoring parent can be rejected by an unresponsive child. An affectionate child can be brushed aside by an uncaring parent. Partners may have one-sided relationships in which a great outpouring of love from one is met with nothing more than cool politeness from the other. A loyal and devoted employee with huge affection and admiration for her employer may, callously, be made redundant after decades of faithful service. Free will is capable of far worse choices than those: it frequently selects war, assassination, murder, theft, betrayal, and treachery as means to selfish ends.

The next theological dilemma is the attempt to discover the balance between the benign qualities of free will — which alone makes love possible — and the malign qualities, which all too often lead to pain, suffering, grief, and death. There is also the problem of order and consistency within the cosmos. If nature was totally unpredictable and capricious, if up was down on Tuesday and yellow was green on Friday, if lions were amiable herbivores on Monday and savaged antelopes on Thursday, there could be no understanding and no learning process. There must be consistency in the universe if our God-given intelligence and reasoning powers are to be our means of making progress. Theologians can readily accept that if God constantly intervened — by performing miracles with time, for example — there could be no real freedom of choice. This knotty question of free will and its theoretical limits can be illustrated by the example of the murderous thief. If God intervenes in the thief's mind and makes him gentle, kind, and good, so that he no longer wants to kill people, the thief has lost his freedom. If God miraculously mistimes the thief's shot at his victim, then God has again intervened. If the bullet drops harmlessly to the grass instead of entering the victim's heart, God has interfered with the laws of physics. If God renders the victim impervious to high-velocity lead projectiles, then God has interfered with the laws of biology. If God prompts the potential victim to travel by another, safer route, then God has again taken away free will. There are no quick or easy answers. The enigma of a totally benign and loving God in a universe filled with suffering is a jagged rock on which many a theological ship has come to grief. An intellectual lifeboat can be constructed from the concepts of love's utter dependence upon free will, and the need for consistency in the laws of nature. This lifeboat is better than nothing — but it is a frail craft with which to challenge a vast ocean of philosophical and theological questions. If, as we believe, our omnipotent, omniscient, and omnipresent God of love is the absolute master of time, why is there so much tragedy here? The theological key to that box of riddles may lie in the mysterious nature of time itself — and in the hereafter, when our earthly time perspective is changed by death. What seems like unspeakable and irreversible tragedy and disaster in the here and now may well be repairable in the world to come. It falls well within the power of omnipotence to put things right in another, future realm, without depriving us of our essential free will in this one: and God's power over time may well be part of that loving compensatory process.

In this chapter, we have looked at some of the unanswerable questions that theologians wrestle with, and in particular, we have focused on the theology of time. In Chapter 7 we examine the mysteries of reincarnation and déjà vu and their relevance to a wider and deeper understanding of the mysteries and secrets of time.

MYSTERIES OF REINCARNATION
AND DÉJÀ VU

THERE ARE various mysterious phenomena that can be examined under the general heading of reincarnation, and the topic is an important one for inclusion among the mysteries and secrets of time. If reincarnation is a genuine, objective fact, it is likely to influence our perspective on the nature and meaning of time. Reincarnation suggests that the immortal essence of the person — the eternal part that is repeatedly reincarnated — seems to be capable of stepping out of time at death and re-entering it at the next birth. This raises a fascinating question: where does it go between its terrestrial lives? Are there other planes of existence? Or does it rest and wait in some timeless zone, a kind of limbo (but not in the theological sense of the word)?

The Hindu faith is arguably one of the best places to start an in-depth study of reincarnation, because belief in reincarnation is an essential part of Hinduism. Their doctrine posits that every living entity has at its core something eternal, something that has always existed and always will exist. This belief can be linked with the ideas discussed earlier about the concept of eternal, infinite time versus the concept of limited time.

An eternal spirit, in the Hindu sense, needs to inhabit an eternal time. Their basic belief is that every living being consists of just such an eternal essence, clothed in flesh during its life on Earth. When physical death occurs, the spirit is released from the flesh and passes into another body. Does this necessarily happen immediately? Different reincarnation theories suggest different answers. The type of body in which the immortal spirit is reincarnated (literally "made flesh again") is thought to be governed by the laws of karma.

The word *karma* itself is taken from an ancient Sanskrit root associated with the idea of action and performing deeds — and with the inexorable consequences of such deeds. Karma, therefore, is about the *results* of actions, the *effects* of actions. Karma is regarded as the sum total of every action: those done in the past, those being undertaken now, and those that will be undertaken in the future. This sense of karma, involving future actions, again raises profound philosophical and theological questions concerning the likelihood of concrete and invariable predestination versus the bifurcated concept of time that incorporates probability tracks and human free will. The laws of karma suggest that we *do* have free will, and, as an autonomous being, each individual goes through life making free and conscious choices.

Yet karmic theology is rather more subtle and complex than that: it suggests that what we have chosen as part of the exercise of our free will actually *creates* the situations inside which we have to make the essential choices. It is also important to remember that the theory of karma incorporates nothing judgmental. Evil is not punished; neither is goodness rewarded. Karma is not retributive. It is believed to be the deeds themselves that create karmic experiences — and that's what makes us responsible for our own lives. Reincarnation theorists who accept the laws of karma would argue that the working of those laws extends through many lives: actions taken centuries ago — during previous incarnations — may still be working themselves out in our present twenty-first-century incarnation.

When some early esoteric Christian mystics adopted the laws of karma, they modified them to take the love, mercy, and forgiveness of God into account. What might, therefore, be described as the law of mystical Christian karma maintains that God's loving intervention and forgiveness operate together to restore our human immortality and destroy the powers of sin and death. A negative karmic experience in a previous life, which carries negative outworkings with it, can — according to these teachings — be cancelled out by benign divine intervention.

Another aspect of reincarnation theory is the "flowing together" concept, or *samsara*, referred to in the sacred writings known as the Upanishads. Our past lives, this present life, and our future lives are not separate: they simply flow together as a *samsaran* totality. Many Indian philosophical and religious traditions embrace the concept of a wheel of life — an ever-rolling cycle of birth, death, and rebirth. This idea is found in Buddhism, Hinduism, and Jainism.

There is a tendency among some Indian schools of thought to regard samsara as a thing to be avoided: an inferior, negative state of being from which it is prudent to seek escape. Other Indian philosophers regard *samsara* as an illusion. To them our lives, deaths, and rebirths aren't real: only the eternal realm is real. This philosophy teaches that it is human ignorance which allows us to think of this world, and our part in it, as real.

The Upanishads form part of the Vedas, the sacred Hindu scriptures; they are also known as the Vedânta ("end of the Vedas"). Most experts date them from the eighth to the fourth century b.c. Unlike the claims made by many "revealed" religions concerning the origins of their respective holy books, some Hindu philosophers use the term apaurusheya when referring to the origins of the Vedas. *Apaurusheya* translates as "without an author," and these Hindu sages argue that the Vedas were written neither by the gods nor by men. Hindu scholars see them as somehow just happening — being both natural and eternal.

Some reincarnation theorists believe that reincarnation is accomplished when the immortal essence — the non-physical soul or spirit, the disembodied mind of the entity — leaves a physical body that is no longer functional and enters a new one. What evidence is there for the existence of such a non-physical entity?

Philosophers, theologians, medical scientists, and psychologists are all concerned with the so-called mind-brain problem.

There are some theorists who believe that what we think of as *mind* is not much more than an illusion, and that our self-conscious belief that we are sentient entities may be, at worst, no more than a trivial and meaningless side effect of biological and electrochemical processes going on within the physical brain. In their view the highly complex physical brain is extremely busy running the body and doing all it can to preserve both the individual and the species. Because this electro-chemical brain is so complex, it produces illusory ideas about its own selfhood and so-called personality, which may, perhaps, assist the organism to survive long enough to reproduce.

Other thinkers would argue that the mind is very much a separate entity, and that it is far more powerful and significant than the brain — a device that the mind steers, directs, and controls. This suggests the analogy of an airplane in which the mind is the pilot, the physical brain is the joystick and control panel, and the body is the engine, propeller, fuselage, wings, tail, and landing gear.

Take the philosophical end of the problem first: we are looking for the nexus between the regular, day-to-day activities of the physical brain (control of breathing, blood pressure, digestion, etc.) and what might best be described as our introspective experiences (investigating unsolved mysteries, playing chess, undertaking post-graduate research, falling in love, enjoying TV, radio, or a book).

There is undeniably an inner, mental life. (Descartes said, "I *think*, therefore I *am*.") This inner life is regarded as undeniable because we all *think* that we experience it, and we share our thoughts about it with other beings whom we believe to be sentient entities like ourselves. This inner mental life is personal. It is subjective. We want certain things. We do not want other things. We plan our lives accordingly so that we can maximize happy, positive experiences and avoid unpleasant, negative experiences. We are looking here — in our examination of our minds — at exactly who or what it is (the *I* and the *me*) that is self-aware, and is doing all this thinking, planning, feeling, and experiencing.

Superficially at least there would seem to be three fundamental possibilities.

The first states that what we think of as private, personal, subjective thoughts and feelings simply do not exist. What we call introspective data is illusory. The only things that exist are the objective, external ones. The brain (simply a sophisticated physical, electrochemical, and biological organ) carries out information processing, just as a robot or a computer would do. There is no mind-brain problem because mind doesn't exist. (The authors are not impressed by this argument!)

The second proposes that all that we think of as subjective experiences, all our cognitive and affective ideas and feelings, *do* enjoy a genuine reality and a bona fide existence independent of the physical brain. Mind, soul, or spirit may be insubstantial in any material, chemical or physical sense. It may have no measurable mass and occupy no spatial volume — yet it most certainly *exists*, and, in the final analysis, is infinitely more important than the synaptic processes that characterize normal brain activity. (This comes fairly close to the authors' own view of the mind-brain relationship.)

The third theorizes that incorporeal mind events and physical brain events are both real, and they interact continually. It is the determined mind, setting out on a life-and-death mission, that tunes and energizes the physical brain to ever greater output, and hence to the success of the mind-directed mission. It is the physical brain (responding automatically to internal pain and exhaustion stimuli and continual tiredness messages from various parts of the body) that tries to persuade the mind to take a rest — and so jeopardizes the mission. The mind here can be seen as the dedicated driver of a Formula One racing car and the brain as the steering wheel and control panel. The tires, brake linings, transmission gear, and fuel tank all send signals to the control panel. This is an especially relevant parallel in the case of luxury saloon cars equipped with talking computers that tell the driver when safety straps are unbuckled, the lights are on, the engine is overheating, or the transmission has failed in some way and dropped down into emergency mode. Everything depends upon the will and the skill of the driver — the *mind*. The driver in turn is dependent upon the correct and responsive functioning of the controls.

C.S. Lewis writing in *The Abolition of Man* argues that the mind's control over brain and body works in two stages. In Lewis's analogy, our powerful basic instincts are visceral, and our abstract minds seek to control them without too much success — but in between our higher, logical, intellectual selves and our more powerful visceral instincts are what Lewis calls our *thoracic* selves: the place where our hearts are found in the emotional, symbolic sense. Lewis's argument is that we can only overcome our strong, basic instincts by allying the mind with the best of our emotions. There is a shipwreck, for example, and a powerful swimmer's visceral instinct is to swim to safety regardless of anyone else. But his ideals — his nobler emotions — tell him that he *must* rescue a non-swimmer nearby, who will certainly drown without his help. Inspired by these thoracic ideals of altruism and heroism, he overcomes his basic, visceral instinct of self-preservation and risks his own life to save the non-swimmer.

Whether Lewis's formidable intellect is correct on this occasion, he certainly seems to be supporting the case for the separate, independent existence of non-physical mind — and if it *does* exist (a theory that the authors strongly support) may it not be capable of moving from body to body as the centuries roll?

Hypno-regression, in the hope of recalling past lives, is one avenue of investigation into the mysteries of reincarnation that co-author Lionel has studied personally and in considerable depth.

There would seem to be three possible alternative origins for what appear to be memories of past incarnations. The first is that they are what psychologists and psychiatrists categorize as subordinate, or repressed, personalities. These repressed personalities can be described as the people we might have been or might have wished to be. When everyday life seems boring, dull, and unadventurous, it may occur to us way down in the depths of the subconscious that being a kidnapped princess awaiting rescue from the ogre's cave or dragon's den would be an exciting alternative. The suppressed vision of being a gallant knight on his great white charger may be a subconscious response to picking up litter in the park, driving a bus, or filling the shelves in an urban supermarket. These repressed personalities can perhaps emerge with a little help from a good hypno-regression expert.

The second theory involves telepathy. One of the witnesses, or perhaps the hypno-regressionist, is *thinking* of who the subject might have been. Was this twenty-first-century, two-hundred-plus-pound weight trainer and martial arts instructor a prizefighter in Regency England during a previous life? Was he a pirate, a highwayman, a Bow Street Runner, or a Victorian pioneer cutting his way through unexplored African jungles? Looking at him in his twenty-first-century body, it would be hard to think of him as a shepherd, a ploughman, or a studious monk writing away conscientiously in a quiet scriptorium.

Co-author Lionel with expert hypno-regressionist Rosie Malone.

The third theory fits those occasions when the so-called past-life memories are historically incorrect. The wrong king or queen is on the throne. The technology doesn't fit the period correctly. The characters the subject thinks he met belong to another place and another time. For example, during co-author Lionel's hypno-regression experiments with Rosie Malone, he "remembered" a previous incarnation as an infantryman at the Battle of Waterloo. This soldier, whose name was George Haskins, described in detail how he had loaded and used his musket. However, he called it a wheel-lock. That created problems. The wheel-lock firing mechanism for muskets and pistols was first developed *circa* 1500. It was replaced in the mid-sixteenth century with the snaphance, and that was superseded by the flintlock around 1600. It is extremely unlikely — although not impossible — that George would have gone into battle in 1812 using a wheel-lock musket. His might just have been an odd case in which a soldier had brought an older weapon with him — perhaps a family heirloom — as well as his early nineteenth-century Brown Bess flintlock, the regulation issue. Was he just taking sensible precautions by carrying a spare musket? Before dismissing the details of that past life as errors of such magnitude that they discredit the theory of reincarnation altogether, let's go back to the idea of probability tracks. Is it conceivable that reincarnation is independent of the single, unidirectional timeline that we think of as history? Can an incarnation on probability track 279, for example, reappear on probability track 432? In the case of George Haskins's anachronistic musket, had his life been lived on an early nineteenth-century probability track in which the wheel-lock had not been replaced by snaphances and flintlocks?

The hypno-regression experiments that co-author Lionel carried out with Rosie Malone were carefully recorded on Sunday, June 18, 2000, and transcribed the next day, Monday, June 19. They were carried out in conjunction with *Fortean Studies*, and later published in Volume Seven of that series, the copyright remaining with the authors. The value of this 2001 *Fortean Studies* publication is that it provides near-irrefutable documentary confirmation of the

Co-author Lionel undergoing hypno-regression.

regressions, as they were recorded immediately after the experiments. The all-important details, therefore, do not depend upon the authors' own typescripts or on their memories of the events.

Rosie began by taking Lionel back to when he was a ten-year-old schoolboy. She then regressed him to a time before he was born in this life. Rosie, as an accomplished professional hypno-regressionist, emphasized that the subject should stop only at points in the past where he felt safe and comfortable, and in no danger of trauma. The choice of where he stopped was, therefore, Lionel's. Rosie also explained that when he had reached such a point, he should open his eyes to signal to her that he was ready to proceed. The George Haskins experience follows:

Rosie:	My name's Rosie, what's yours?
Lionel:	George.
Rosie:	Just George?
Lionel:	George Haskins.
Rosie:	Where are we, George?
Lionel:	Getting ready for battle.
Rosie:	Do you know where we are? I don't recognize anything here.
Lionel:	There's a big field. There are tents everywhere.
Rosie:	Do you know the name of the place?
Lionel:	We are in France. I don't know the name. I can't pronounce it.
Rosie:	Are you in the army?
Lionel:	Yeah.
Rosie:	What do you do in the army?
Lionel:	I'm an infantryman.
Rosie:	What does that mean?
Lionel:	Carrying a musket in the front line.
Rosie:	Oh, right.
Lionel:	It's a wheel-lock.
Rosie:	A what?
Lionel:	It's a wheel-lock.
Rosie:	A wheel-lock?

Lionel: A musket. You load it with a rod from the front.

Rosie: Oh, right.

Lionel: You carry powder and shot in a little pouch at your belt. You've got enough for ten rounds.

Rosie: What do you do then?

Lionel: Wait for the Quartermaster.

Rosie: And he brings you some more, does he?

Lionel: Sends his men. Don't come himself!

Rosie: Oh no, he's too important?

Lionel: He's back of the line. He sends a runner with the shot.

Rosie: What's the name of the army that you're in?

Lionel: Wellington.

Rosie: Wellington? Oh, the Iron Duke!

Lionel: British. We're British. That's what we call him.

Rosie: The Iron Duke?

Lionel: Arthur. Arthur something.

Rosie: Oh yes.

Lionel: The Iron Duke we call him.

Rosie: The Iron Duke you call him. Yes, that's what I've heard him called. Have you ever seen him?

Lionel: On a big white horse.

Rosie: Have you?

Lionel: Big hat, lot of brocade, cloak, glittering in the sun.

Rosie: Were you close to him?

Lionel: I heard him speak. I could have reached out and touched him.

Rosie: Oh, that must have been great.

Lionel: He rode down the line talking to us.

Rosie: And you're fighting the French?

Lionel: Fighting the French — Boney!

Rosie: Boney? Oh, that's Napoleon Bonaparte, isn't it?

Lionel: Yeah. Boney we call him — old Boney.

Rosie: Oh, right. Have you ever seen him?

Lionel: Only pictures of him.

Rosie: Aha, yeah.

Lionel: Haven't seen him. Boys draw pictures of him.

Rosie: Yeah, I bet they do.

Lionel: Got his hand stuck in his weskit.

Rosie: Ah, right. Do you get paid?

Lionel: Shilling a day.

Rosie: Shilling a day?

Lionel: While we're active — shilling a day.

Rosie: What about when you're not active?

Lionel: Three shillings and sixpence for the week when we're not active.

Rosie: Oh, right. When you're not fighting in France, where do you live in England?

Lionel: Dorset.

Rosie: Dorset? Oh, it's nice down there, isn't it? The name of the place in Dorset? I know some of the places.

Lionel: Tovey.

Rosie: Where?

Lionel: Tovey.

Rosie: Tovey. Are you married, George?

Lionel: No, single boy.

Rosie: Are you? Ah.

Lionel: I was going to get married. Then I joined the army.

Rosie: Oh, right.

Lionel: 'Cos that was better pay.

Rosie: Yes, it's good pay. Is it dangerous?

Lionel: Good friends killed … too many, too many.

Rosie: It's not easy, is it?

Lionel: That's come close once or twice.

Rosie: It would do, wouldn't it? Now, George, I want you to go back to sleep again …

Just as there were interesting queries about George's anachronistic wheel-lock in 1812, there are problems with the village of Tovey in Dorset. Searching a gazetteer carefully shows no sign of it, but there is a memorial brass on the wall of Langton Matravers church in Dorset, honouring the fearless Admiral Tovey. A search of a *world* gazetteer, however, revealed that there is a small township named Tovey in Illinois, United States, with a population of less than a thousand at the most recent census. This Tovey is situated at latitude 39.588 north, longitude 89.456 west.

As far as the imagination can stretch around infinite probability tracks, there doesn't seem to be any comprehensible nexus between George Haskins's Dorset village of Tovey (which does not appear to exist in our world), the township of Tovey in Illinois, and a heroic and much-loved admiral bearing that honourable name who did come from Dorset!

The next character who emerged from this hypno-regression experiment was Peter Walters, a stonemason's labourer.

Lionel: I don't do the fine bits.

Rosie: You're learning, are you?

Lionel: I carry the stone.

Rosie: How old are you, Peter?

Lionel: I'm not rightly sure. Fifty-something they tell me.

Rosie: Did you have any schooling, Peter?

Lionel: No, only what Mum and Dad taught me at home. I don't know my letters.

Rosie: Where do you live, Peter?

Lionel: Peterborough — that's why they called me Peter.

Rosie: That's lovely. What's your Mum and Dad's name, Peter? Do you know?

Lionel: They're dead now.

Rosie: Yeah, but when they were with you what were they...

Lionel [interrupting her]: Henry was my father.

Rosie: Henry who?

Lionel: Walters, Henry Walters.

Rosie: So you're Peter Walters, are you?

Lionel: They call me Peter the Mason. I'm Peter Walters really, but I can't write it.

Rosie: That's not necessary.

Lionel: I can make my mark.

Rosie: Where are we? I don't recognize anything.

Lionel: We're in the quarry.

Rosie: Where is this quarry?

Lionel: That's over the border. It's a long way from where I live. We're in Wales somewhere — over the border — where we buy the stone. Master buys the stone and he gets me to come with him, 'cos I hump it about. He looks after me all right. He's nice.

Rosie: Does he pay you?

Lionel: I get a groat a day and he feeds me. That's all right. That's all I need. I've got to save up 'cos I'm getting on now. I've got several, in a jar, put away — in a stone jar.

Rosie: Look at these groats — which king or queen's head is on them?

Lionel: That's George. That's the only one. That's George and he comes from Germany.

Rosie: He comes from Hanover.

Lionel: Comes from Germany, so master tells me. We'd gone up to London to do some work… Westminster. We were repairing things. I was humping the stone, and they all said, "Stand still — that's a royal procession." We all had to stop hammering while it went past. They say she'd come to look at the Abbey. There was a piece of stone that had come down in a storm, like one of the buttresses that had been damaged. I don't know if that was lightning, but that was damaged. We didn't see it damaged; we got there to repair it, and that was all sort of shattered as if that had been hit. I thought at first that it had been hit by a cannon ball when I saw it, but they wouldn't have done nothing like that in the Abbey. He can figure it and he measured up and everything and told me what stone to bring him. I do a little and he let me do the rough ashlar. Rough ashlar, that's the rough cut stone … He draws the lines on it with a piece of chalk and I mustn't go past them in case I make a mistake. Then I take it to him when I got nearly up to the chalk and he say, "That's all right." He sometimes say, "Good boy, Peter," if I can get two of them done in a day. Then he perhaps buy me an extra drink on top of my groat ... That's hard work for them horses up the hill with stone in ... We have to get out sometimes and put our shoulders behind it. We carry this little ... that's one of my jobs. I carry the chock on a rope at the back, and if we're on a hill I have to put the chock under quick 'cos I don't want to hurt the horses and pull them back ... There's Prince and Gelder.

King George I of Great Britain was George Ludwig, Elector of Hanover. He reigned in Britain from August 1, 1714, until his death on June 11, 1727. Peter Walters and his master would have been working on repairs to Westminster Abbey between those dates. It is particularly significant that the two western towers were built by Nicholas Hawksmoor between the years 1722 and 1745, so a number of master masons and labourers would definitely have been working on the site when Peter said he was there. Peter *Walters* said that he and his master had gone over the Welsh border in order to buy first-class stone for their work at Westminster Abbey. It may be only a simple coincidence, but some important Welsh stone quarries currently belong to the prestigious *Walters* Group, who carry out civil engineering and a wide range of associated work. Their headquarters are in Hirwaun, near Aberdare, in Wales.

Peter's comments on the groat are also interesting. After 1662, there was a milled issue of the groat, a four-penny coin. These continued in circulation until the reign of George III (1760–1801). The coin was certainly in circulation during Peter the mason's lifetime.

The next hypno-regression experiment took Lionel back to the days of Brother John, a Franciscan friar.

Lionel: Brother John... the Order of St. Francis. We are near London in a town called Reading, we're near Reading.

Rosie: Do you prefer to be called "Sir" or do you like to be called "Brother John"?

Lionel: I would rather be the brother of all men.

Rosie: So I can call you Brother John?

Lionel: Please do, little daughter.

Rosie: Where do you come from, Brother John?

Lionel: I was born to the far north, in Durham ... in Durham town, by the great cathedral. I have some simple skill at letters, yes ... I can read the classical tongues.

Rosie: And what would they be?

Lionel: Greek and Latin and Hebrew ... and Hebrew, dear child, yes ... it is Edward ... the only Edward I know of. We are in the year of our Lord 1191.

Rosie: When did you enter the religious life?

Lionel: When I was a young man, some forty years ago. We follow the Order of St. Francis, Francis of Assisi.

Rosie: What's the holy rule of St. Francis of Assisi?

Lionel: The holy rule is the love of God and of all our brothers and sisters — the rule of poverty, chastity, and obedience. We have nothing for ourselves. All things are for the church and for the poor.

Rosie: This is the Catholic, the holy Roman Church, is it?

Lionel: This is the Church.

Rosie: Yes, the holy Roman Church. I am an ignorant girl.

Lionel: There is but the great Church of Rome — and those of the East whom we do not commune with.

Rosie: They are the infidel though, aren't they?

Lionel: Those of the East are Byzantium ... It was the Great Schism. Filioque ... There was a difference of opinion among the learned doctors. There were those who said, as we did, that the Holy Spirit proceeded from the Father and the Son. Filioque in the Latin tongue is "and the Son." Those of the East said that the Holy Spirit proceeded only from the Father ... and so the Schism occurred.

Rosie: And will you be in the religious life until your time is ended?

Lionel: Until my time on Earth is ended and I go to be with God, yes ... I help the lay brothers who come and work in the fields and I try to heal them if I can; but most of my time and the work I love best is the copying and translating of the Scriptures.

Rosie: Oh, you do those nice pictures in the Scriptures?

Lionel: We draw and colour in the margins to the glory of God, so that those who are shown the books and have difficulty with their letters can understand something of the pictures ... they also serve when we are reading from the books as index marks. When it is a brother's turn to read and he is turning the holy pages it is easier to see the colour and the height from the top of the page ... thus we find the page quickly and do not delay the service.

Brother John the Franciscan poses several problems in terms of his dates. He said that he was living in 1191, and that he had been a Franciscan friar for forty years, which would have taken him back to being a young man in 1151. St. Francis of Assisi was born in 1181 and died in 1226. He established his order during the first decade of the thirteenth century. The future St. Francis was only ten years old when Brother John said he belonged to the Franciscan order.

Brother John also creates problems when he refers to King Edward. Edward the Elder, of the House of Wessex, reigned from 899 to 925, which was two centuries too early. Edward the Martyr, also of the House of Wessex, reigned from 975 to 978, also much too early. The Saxon monarch, Edward the Confessor, reigned from 1042 to 1066: still a century and a half ahead of Brother John's date of 1191. The Plantagenet Edwards I, II, and III reigned from 1272 to 1307, 1307 to 1327, and 1327 to 1377 respectively. All of them were too late in time to be Brother John's Edward. It was Richard the Lionheart who was on the throne from 1189 to 1199. It may seem like clutching at straws on behalf of reincarnation theory, but it is worth looking again at what Brother John actually *said*: "... it is Edward ... the only Edward I know of ..."

A sincerely religious and well-educated man like John would have known (as part of his studies) of the historical existence of Edward the Elder, Edward the Martyr, and Edward the Confessor. When Brother

John uses the expression "it is Edward ... the only Edward I know of," could he possibly mean *the only Edward I recognize* or *the only Edward I acknowledge as a real and worthwhile king, because of his spirituality*? Was he then referring to the Confessor, or the Martyr? However, both these kings still leave us too early.

His Franciscan dating seems to be an intractable problem: but what if John had said 1291 instead of 1191? Just suppose there had been a typing error — and in the best and most careful write-ups of experiments and research work, typos do occur occasionally. The change of a single digit from 1 to 2 is understandable; after all, they are side by side on the keyboard. If it ought to have been 1291 instead of the 1191 that we recorded immediately after the experiments with Rosie, then not only is the Franciscan date problem solved — the King Edward problem is solved too! The Franciscan Order was well known and firmly established by 1291, and England was ruled by the powerful Edward I (known as Longshanks). Such were his fame and military conquests that Brother John might well have said "it is Edward ... the only Edward I know of."

Richard the Lionheart.

Brother John's reference to Durham and its great cathedral fits well with actual history. The cathedral was founded in 1093 and was also known as the Cathedral Church of Christ, Blessed Mary the Virgin, and St. Cuthbert of Durham. The relics of St. Cuthbert of Lindisfarne are kept there along with the head of St. Oswald of Northumbria and the remains of the Venerable Bede. It is exactly the kind of awe-inspiring religious centre that would have influenced a Durham boy like John towards the religious life.

John's comments on the Great Schism and the arguments about

the Filioque clause are also historically correct. If John lived in either
the twelfth or thirteenth centuries
— and the thirteenth seems more
likely — he would have known all
about the religious tragedy of the
Great Schism that took place on
July 6, 1054.

Durham Cathedral.

The two leading actors in
the tragedy were the Patriarch
of Constantinople, Michael
Cerularius, leader of the Eastern
Church, and Pope Leo IX, leader of
the Western Church. Cerularius
had excommunicated the Bishops
of Constantinople for observing
Western Church practices that he
didn't like. In April 1054, Pope Leo
sent a legation to Cerularius, led by
Cardinal Humbert. Leo's message
consisted of authoritative demands
and numerous accusations levelled
against Patriarch Cerularius. To
complicate matters further, Leo
died in the middle of the dispute,
which meant technically that
Humbert's powers as a legate had
died as well — but Humbert was a
tough and determined character,
so he continued with what he
believed to be the job at hand. He
and Cerularius seem to have disliked each other intensely on sight,
and neither man had enough toler-
ance or diplomacy to fill a thimble.
The Romans excommunicated

The mysterious ancient door knock-
er of Durham Cathedral — does it
indicate a pagan past?

Cerularius, who promptly retaliated by excommunicating them. It could all be regarded as farcically funny and ridiculous — were it not for the tragedy of how far it took the Church from its real God-given purpose and mission in the world: to show love and kindness everywhere, to treat all men and women as brothers and sisters, to care for the poor, to heal the sick, and to shelter the homeless.

Brother John, however, seems to have had a firm and realistic hold on religious priorities: "The holy rule is the love of God and of all our brothers and sisters — the rule of poverty, chastity and obedience. We have nothing for ourselves. All things are for the church and for the poor."

Brother John's references to Reading are also historically interesting: "We are near London in a town called Reading, we're near Reading …"

The earliest traces of the settlement that eventually developed into modern Reading were at the confluence of the River Thames and the River Kennet, and in the eighth century it was known as Readingum. Historians have two explanations for the early name: it may have come from an Anglo-Saxon phrase meaning "the home of Readda's people" or from the Celtic *Rhydd-Inge*, meaning "the ford that crosses the river." Vikings lived there in 871 and it is listed in the Domesday Book in 1086

Ancient Greek merchant ship.

as having some six hundred inhabitants. Medieval pilgrims frequented Reading Abbey, which makes sense of Brother John's words, as does another historical reference to a certain John of Reading, who was the Abbot of Osney in 1235. This suggests that Reading during that period was a sufficiently important religious centre to have produced men with the calibre to become abbots.

The next hypno-regression experiment produced a character calling himself Theostas, an Athenian seaman from the first century A.D.

Rosie: Where are we, Theostas? I'm a bit lost.

Lionel: We are on board the ship and I am very weary … it's all right … I am better than I was…. We had fever on the ship … Athenian merchantman … jars of olive oil and wine.

Rosie: Are they those big jars? The amphora?

Lionel: Tall as a man…. You can just put your arms around one. We stow them touching, with straw between them so they do not break during the voyage when the sea is rough. The jars to the straw … four straws, a man pulls the jar and the comrade beside him puts the straw between that and the next jar, and you must hold the straw until the jar is in place…. If the jars break there is no profit in the voyage and the sickness and the danger and the shipwreck is for nothing … I am the freight master. I supervise. I help my men and I supervise them.

Rosie: And the captain's name, the master's name?

Lionel: Solus … He is the master and the captain. He is a merchant captain. It is his merchandise as well … He buys for himself and he sells for himself: he says he trusts no other.

Rosie: Where does he buy from and sell to?

Lionel: We buy from the hill farmers, the vine-growers, and the wine merchants. In the hills beyond Athens in the Peloponnesus to the south … We take it to the ship. We sail for Rome. We bring oil and wine from Greece to Rome. We return with ornaments, silverware, copperware.

Rosie: Who is Caesar at the moment?

Lionel: Caesar Augustus ... It is the seventh year of Caesar Augustus.

Rosie: Who was Caesar before him?

Lionel: Julius.

Rosie: Oh, he was murdered, wasn't he?

Lionel: They were bad times; it was not safe to sail. After great Caesar died there were wars throughout the Empire. The seas were not safe. There were no Imperial Fleets to stop the pirates ... There were many pirates; they are gone now. In trader Greek we keep the words simple and short. I was twelve summers when I went to the Academy; fourteen summers when I went to sea. I am recovering from the fever. It burnt for three days. I thought I would die. It is better now. They brought me fruit ... Oranges and the juice of lemons mixed with wine and water and it was good.

Rosie: Do you get much fever on board ship?

Lionel: Only if the food is not good. We have salted meat ... pork, beef, and venison. Spiced and herbed — but if the money is short, and the spices are not sufficient, there is poison in the meat, like a vapour, and it gives you the fever ... pains in the stomach and the constant flux. I am recovering ... I shall see port again.

Theostas spoke a lot of historical sense about Greek shipping in the period he described. Although Greek maritime power as such was in decline during the Roman period, it was still viable, and the Romans tended to depend on it. Rome had conquered Macedonia in 186 B.C. and the remainder of Greece in 146 B.C. Syria succumbed to Rome in 65 B.C., and Egypt became more or less Romanized in 31 B.C. That gave Rome dominance over most of the known world — but the Romans did not have the affinity with the sea that the Greek mariners had already enjoyed for centuries by the time Theostas came along.

At one time Roman laws and customs either prevented — or strongly discouraged — Roman citizens from becoming merchants or ship-owners. This meant that Rome needed ships and sailors, and the fearless, seafaring Greeks were happy to oblige. The Roman contribution to the Greek mercantile marine during the time of Theostas included building new ports and improving existing ones, building lighthouses, digging canals, and protecting merchant ships from pirates. Another interesting historical sidelight on Theostas and his reference to Augustus is that Augustus had suggested to the ship-owners in Alexandria that it would be advisable to build larger vessels of up to 1,300 tonnes.

The commercial and legal shipping terminology of the time was not very different from that which is customary today. Charter deeds and documents included the names of the owners of the goods; the owners of the vessels; the displacement weight of the ship; payment dates; types of cargo to be loaded; time allowed for charging and discharging; engagement of necessary crewmen; and a stipulation that cargo must be carried in an undamaged and dry condition. The name of the master also had to be included in the documentation. Theostas emphasizes the importance of keeping the cargo safe and secure, and goes into considerable detail about how the straw had to be inserted between the amphoras. He also names the master, a man called Solus. There is a port called Solunto or Solus on the island of Sicily, which would have been well known to Greek

Noble and fearless, a Templar warrior-priest, like Vernon de Grey of Thetford, England.

sailors in the days of Theostas. In an era when a man was often named after the place where he was born, could Solus, master of the vessel on which Theostas was in charge of the cargo, have been a Sicilian?

If, as Theostas says, Augustus Caesar had been ruling for some seven years, the date would have been *circa* 20 B.C. The Senate gave Octavian full powers and the title Augustus in 27 B.C.

Theostas's reference to the Academy is also interesting. The original Academy was founded by Plato in 387 B.C. and ran with a few minor interruptions until it was closed by the Byzantine Emperor Justinian in 589 A.D. Theostas might well have attended the Academy as he said he had done.

His detailed comments on the fever — possibly a form of food poisoning — also seem reasonable and realistic enough.

The final hypno-regression experiment led to the emergence of a character who called himself Vernon de Grey, a nobleman from East Anglia who joined the Templars after the death of his beloved wife, Elizabeth. Traditionally, there was a famous de Grey family in the Breckland area at one time, and according to folklore they were associated with the legendary tragedy of the Babes in the Wood.

Rosie: Where are we, Vernon? I don't recognize this place.

Lionel: We are in the Holy Land ... I can read and write; sign my name and seal it; yes ... I am a member of the Order. I defend the paths for the pilgrims ... I am of the Order of the Temple, the Poor Knights of Christ. A military order, an order of priests ... All of us in the Order of the Temple are both priest and warrior for Christ ... There are lay brothers who attend us, but all who ride in arms with the Great Master are ordained.

Rosie: Who is the Grand Master at present?

Lionel: Guy.

Rosie: Guy who?

Lionel: Guy of Gervaise ... We meet when we need to on the full moon, or as near as we can ... Regularly, depending upon the Grand Master's travels and upon the war with the Saracens. It is good to meet; it is better to fight. We are few and they are many, but we must hold the roads for the pilgrims.

Rosie: Which particular road are you holding at the moment?

Lionel: Jerusalem to Damascus.

Rosie: Which shrine is it they are going to worship in?

Lionel: Damascus — there is a shrine along the road sacred to St. Paul, where he saw the vision of Christ ... many shrines ... the Holy Sepulchre, especially the Holy Sepulchre ... I've been here four years.

Rosie: And where were you before you came to the Holy Land?

Lionel: I was at home ... near to Thetford, in the land of the East Angles in England.

Rosie: Are you married?

Lionel: I had a lady at home ... Elizabeth ... She is gone ... She died of the plague.

Rosie: As a priest, are you allowed to be married?

Lionel: I joined the Order when she died ... None of us in the Order are married. We worship together at our Templar meetings, but it is not a public worship. We will join public worship, if we are asked. But our own Holy Meetings, by the Secret Order of the Temple, are ours alone ... When Elizabeth died, I wanted to die — I do not know how the plague took her and missed me. We were always together. Then I heard a voice that said it would be wrong to die by my own hand, and if I did I would not see her again in God's Holy Kingdom. So I sought death by the next best means: I joined the Order. I rode to Jerusalem. Sold my estates ... I enjoy hunting for Death ... I enjoy battle ... I cannot find Death. It eludes me ... I ride at the front and seek the thick of the battle. No arrow finds me. No sword pierces my shield. I have slain fifty, perhaps one hundred of the enemy. All around me ... and still I came through their battle lines ... I don't know how.

Rosie: Maybe that's God's will.

Lionel: It must be; there is no other explanation. I have sought death so hard.

Rosie: Maybe you shouldn't search so hard.

Lionel: I want to be with Elizabeth ... There are brigands here as well, coming from the mountains on either side of the road in the wilderness. They hide out in the caves and descend on the pilgrims, but we have our own games. As they hide, so we hide. We found, but three nights ago, a brigands' cave — seven of us, and we set a plan, we set a trap. We let the next small group of unarmed pilgrims pass the road, and then we lay in wait for the brigands ... as they were about to ride down on the pilgrims so we rode down on them ... and it was wonderful. Not one escaped, not one. They will trouble that road no more. But we gave them Christian burial ... We gave them fair fight ... We never attack from the rear. Always from the front ... A frightened man with a crossbow doesn't shoot straight ... Once we had closed with them it was all over ... We buried nine of them. We gave their horses to the pilgrims to sell in Damascus.

Rosie: Do you have a name for your horse?

Lionel: Delys ... I call my horse Delys ... My estates were in Breckland near Thetford, in the land of the East Angles. We came with Duke William, my ancestors came with Duke William. He gave us the land ... My father taught me to fight ... and my grandfather. The Normans ...We are swordsmen ... With the hilt of a sword when they least expect you to strike back ...With three men round you ... and the treacherous coward who tries to take you from behind ... and you see him in your shield ... and he does not know that you have seen him ... and then the hilt comes back ...

Rosie: The hilt?

Lionel: Up! You have two men in front, and the traitor comes behind ... I polish the inside of my shield as a mirror, as a glass, and I bring back the hilt ... and I crack his skull ... You can use a sword in three ways ... Thrust and stab, and the forward blow to give you best distance. To fall upon one knee and stretch the arm and reach twice the length your foeman thinks you can reach. You take him below the chain mail in the great blood vessel of the leg ... and he falls and dies ... There is the slashing blow which takes off a man's head if you do it right ... It is difficult to change direction ... The armourer will make you the sword you wish for. The light sword has the advantage of a swift turn and manoeuvre, but its blow is not so telling. The heavier sword, which will slice through chain mail, will split the joint of armour. That sword cannot be turned, and

Templar medallion showing two knights sharing a horse.

once you are committed to the blow, it must go through. If you miss you are vulnerable. So there is the slicing blow, the stabbing blow and that favourite trick of mine which is the backward blow of the hilt. I wear the heavy hilt, the armourer made it especially for me. It is not rounded. It is like a pyramid, and it will pierce as well as smash — like the battle-hammers of old.

Rosie: Do you train your horse as well?

Lionel: I fear for my horse. Many of my comrades tell me that I am a fool, that I care too much for the horse. He is trained ... Delys is a warrior with four legs ... He wears armour on the flank and on the head ...

The Templars — as a visible military order — can be dated back to 1118, although their predecessors existed long before that. During the reign of King Baldwin II of the Kingdom of Jerusalem, Hugues de Payens and eight loyal and noble companions in arms offered their services as defenders of the Christian Kingdom. They were quartered in premises adjoining the temple, and so took their title: *pauvres chevaliers du temple* (literally, "the poor knights of the temple"). Rightly renowned for their breathtaking courage and determination, the noble order grew in strength and wealth until it was a force to be reckoned with internationally. None surpassed the Templars in battle or in single combat; nothing could equal their awesome military architecture; few even guessed at the amazing ancient secrets that they guarded so well. Their positive contribution to history was out of all proportion to their meagre numbers. It was one of the greatest tragedies of all time when their noble order was treacherously attacked by the odious Philip IV of France (ironically known as Philip le Bel) on Friday, October 13, 1307 — but his success was only partial. Some Templar warriors and their fleet escaped. Vitally important Templar secrets were hidden far beyond the reach of Philip's minions. The great Templar principles have survived into our own twenty-first century: the Templars themselves live and work on today — often in secret — for the good of all humanity. They still fulfil their major role as powerful protectors of all that is good.

So we can fit Vernon de Grey of Breckland into the period between 1118 and 1307: a date near 1280 or 1290 would probably be a reasonable one. There was an English leader of the Templars, Guy de Foresta, who was in charge during the last decade of the thirteenth century. Was it *this* Guy whom Vernon meant? It is possible that Gervaise was a nickname of some kind, rather than a proper surname. Gervaise derives from the Latinized form Gervasius, from an earlier Germanic root meaning "spear." Was Guy de Foresta respected for his spear-throwing powers?

Vernon is clearly a committed, dedicated, and absolutely determined man expressing very powerful emotions. Heartbroken after losing Elizabeth, he pursues death as a hungry lion pursues a zebra. He is also bold, ruthless, and proud of his strength and fighting skills — as if his last pleasure in life consists of bringing down his enemies. The details of his mediaeval combat techniques seem realistic and effective. The tantalizing question about his historic accuracy remains.

Co-author Lionel's picture, taken with a Kirlian camera to show the aura.

Can the mysteries of Kirlian photography be connected in any way to the mysteries of reincarnation in general, and to co-author Lionel's hypno-regression experiments with Rosie in particular? Kirlian photography is named after Semyon Kirlian, who discovered in 1939 that a strong energy field — usually electrical — could produce an image on a photographic surface. Do previous lives show themselves in this way? Could the different portions of the so-called aura in the picture be the vestigial traces of earlier incarnations? They could equally well, perhaps, be subordinate, repressed personalities. Or might they be literary creations waiting to be written into films, novels, poems, or plays? Are these coloured clouds of energy the early stages of creating a tulpa, of the kind that Alexandra David-Neel described in *Magic and Mystery in Tibet*? By concentrating and performing the necessary rites, Alexandra reported creating a monk-like tulpa that later escaped from control and reportedly took six months of concentrated mental effort to dissolve. The most interesting and significant aspect of a tulpa phenomenon like Alexandra's monk is that others can see it as well!

As co-author Lionel is a professional TV presenter and radio broadcaster, in addition to his own hypno-regression experiments with Rosie Malone he worked with the very popular British TV breakfast series *Good Morning* in the role of their "regression detective." In order to safeguard the identities and personal privacy of the celebrity subjects who volunteered to take part in this outstandingly interesting series of televised regressions, it is possible to give only a brief outline of their cases. An excellent professional hypno-regression expert carried out these sessions, the highlights of which were then included in the TV programs. The first subject, whom we shall refer to as Mr. A, recalled a previous life in which he had been living in Croydon during the Second World War and taking part in the planning of Operation Overlord, which was the master plan behind the D-Day landings.

These planning meetings were top secret and were held in what was then the Victoria Hotel (now the Nigerian Embassy) in London. These Operation Overlord conclaves were so secret that everything surrounding them was strictly on a need-to-know basis. In this apparent previous life in the 1940s, Mr. A was responsible for taking the minutes at these top-secret meetings.

When carrying out his regression detective role in connection with this case, Lionel visited the national archives and found that minutes of these highly confidential meetings had certainly been taken by the character whom Mr. A thought he had been in the 1940s. The significant factor in this case is that the character who appeared during the regression had *known* that secret meetings regarding Operation Overlord had been held in the Victoria Hotel at that vitally important time — and how could he have known, unless he was actually there?

Mr. B recalled a previous incarnation as Jacob Smithe, or Smythe, in Ampleford, where he had lived in a stately home called Grantham House. In particular he commented on the numerous fine windows that it possessed. He also described his career as a surgeon and medical lecturer at St. George's Hospital in London, and the tragic death from tuberculosis of his beautiful young wife, Emily, when she was only twenty years old. In his role as the regression detective, Lionel found that both Grantham House and St. George's Hospital were real places, and

St. Joachim's Thaler — an unusual early coin.

that surgical lectures had taken place in that teaching hospital early in the nineteenth century, when Jacob said that he worked there in that capacity. It was also significant that attractive windows were a prominent feature of luxurious Grantham House.

Ms. C recalled a previous life as Ida Rowan, who had given up her two children to be brought up in an orphanage in Manchester. Lionel's work as the regression detective led to a former orphanage situated in Little Nelson Street, known in the early twentieth century as the Ragged School and Working Girls' Home — the place to which the two children, Debbie Anne and Beth, had apparently been sent when Ida gave them up. Unfortunately, all the records were inaccessible. Further regression detection work revealed that an Ida Rowan had been included in the 1901 census. Subsequent investigation led to addresses connected with her in Wright Street and Cole Street — frustratingly, both houses had been demolished, but it was possible to visit the sites where the streets had once been.

Mr. D provided the most significant and challenging regression of all. He recalled his life in the sixteenth century, as a young Dutchman who had walked all the way to Bohemia and found work there as a silver miner. Through strength and determination, he had worked his way to great wealth founded on silver, and was involved in the minting of coins known as St. Joachim's Thalers. The silver for these coins, which were made from 1518 onwards, came from a mine near Joachminsthal ("the valley of St. Joachim," traditionally the husband of St. Anne and father of St. Mary the Virgin). Joachim's image was featured on the coin.

Lionel's regression detective work took him to the British Museum, where the Curator of Coins and Medals very kindly and helpfully produced a sample of the coin. Mr. D's regression had produced the right time, the right location, and the right coin. It seemed to be far more than simple coincidence.

From Lionel's personal first-hand experiences and investigations in this chapter, we move on in the next chapter to examine many more reports of regression and déjà vu.

Chapter 8

MORE MYSTERIES OF REINCARNATION, TIME SLIPS, AND DÉJÀ VU

AMONG THE many fascinating phenomena that impinge upon our consideration of the nature of time is telepathy. There are substantial bodies of evidence suggesting that telepathic phenomena are genuine and that they can be independent of distance. If these telepathic communications are instantaneous, can it be argued that in some way thought *transcends* time? We are not referring here to the normal, brain-directed thought processes or reactions that can be timed: the speed with which we withdraw a burnt finger from a hot surface; the attacking and defensive moves of a swordsman or karate expert; the braking and steering manoeuvres of a professional race car driver. What if a telepathic message can cover significant distances in no time at all? Does that not also argue in favour of an immaterial mind that transcends the physical brain, superior to it and independent of it? If telepathic communications can happen instantaneously, what does that say about the nature of time? Does instantaneous telepathy mean that the non-physical mind can get outside time? If it can, then what manner of substance constitutes a time that can be entered and exited at will?

The authors' great friend Robert Snow, a dedicated, experienced, and extremely reliable investigator of the paranormal and anomalous, has shared some of his intriguing telepathic incidents with us.

Robert's friend Jeff R. in Vermont, United States, was diagnosed with terminal cancer, but underwent two operations to slow the progress of the disease and prolong his life as far as possible. Jeff's cousin, Sylvia T., lives in Cheshire, United Kingdom, and kept Robert informed periodically about Jeff's condition, as did Jeff's wife, Elma. Robert had heard from neither of the ladies for some days, and naturally assumed that Jeff's condition was more or less unchanged. However, on the morning of Friday, August 14, 1998, Robert awoke from a particularly vivid and realistic dream in which *someone* — Robert does not know who — informed him that Jeff had just passed over. In the dream, Robert repeatedly asked his unknown informant why Sylvia hadn't informed him. When he awoke from the dream, Robert felt very disturbed and unsettled. He was certain that something was wrong, but as the day wore on this anxious feeling passed, and he began to reassure himself that it was only a dream and that it had no sinister significance. Four days later, he had almost completely dismissed the strange dream from his mind. Then he found a message from Sylvia on his answering machine telling him that Jeff had died. He phoned her and learned that Jeff had left this world at 2:00 a.m. on the morning of Friday, August 14. Allowing for the five-hour time difference, Robert had experienced the weird dream at the precise time that Jeff had passed over.

Sylvia had not been able to inform Robert earlier, as she had been away and had not learned of Jeff's death until four days after it had happened. Robert's own theory, which seems a sensible and valid one, is that Jeff had been thinking of him as he left this world. They had been good and close friends for many years, and bonds of that kind often seem to be conducive to telepathic contact.

Another very moving telepathic episode in Robert's life was associated with the death of his greatly loved mother. Tragically, for the last decade of her life she suffered from Alzheimer's, which gradually worsened year by year. In the later stages of the disease, Robert's mother had deteriorated to the point where she no longer seemed able to recognize anyone, and she also appeared to be incapable of speech. One Sunday morning in the late summer of 2001, when Robert phoned the nursing home, the sister on duty told him that his mother was sinking fast and could last only a matter of hours. Some sensitive and caring staff members had actually advised him against seeing her because the later stages of the illness had emaciated her to the point where she looked almost like a famine victim.

Despite dreading seeing her in that condition, at 1:35 that Sunday afternoon, Robert felt a sudden, undeniable urge to visit her. He drove the fifteen miles to the nursing home where she was being cared for and went straight up to her room. In Robert's own words, it was like a scene from a horror film: his mother was far worse than when he had last seen her. The nurse on duty left them alone together, and Robert took his mother's hand and sat down beside her bed. Tears streamed down her face; she squeezed his hand tenderly, gave him a huge, loving smile, and slid peacefully into the next world. It seemed to Robert that in the moment before her death she regained her lucidity and looked much younger.

How is a case like that of Robert's mother to be understood and explained? In view of recent medical discoveries concerning the retention of conscious lucidity in patients who were otherwise thought of as being in a permanent vegetative state, it would seem reasonable to argue that patients who suffered as Robert's mother did still retained all the awareness and powers of the non-physical mind even though the illness had grievously harmed the physical brain. In her final moments, perhaps, Robert's mother's mind had made a supreme effort and taken charge of her physical brain again — if only to a limited extent. Consequently, she had succeeded first in calling to him telepathically, and then in communicating with him when he was at her bedside. It seems that she was able to let her loving son know that despite what the Alzheimer's had done to her body, her mind and her real personality were still very much alive and intact — and in the eternal and abundant future life they would undoubtedly meet again.

As a professional broadcaster, co-author Lionel is a frequent celebrity guest on George Noory's show *Coast to Coast*, where they discuss various aspects of the paranormal. These broadcasts attract many e-mails from interested listeners who have had strange experiences themselves, and several fascinating accounts of time slips have reached the authors in this way. Three very interesting and intriguing ones come from Toni Keys of New York, who has kindly given us permission to include them. Toni has an attractive and distinct handwriting style, and in 1974 a friend asked Toni to prepare her Yule greeting letter. In Toni's own words:

I was in my office after hours with India ink, my pen, and her letter. As I commenced writing the letter, I found myself writing in a "calligraphy" style (it should be noted: I have never studied calligraphy). I literally got scared because it felt like I wasn't doing the writing. I stopped, looked up, and I was in a stone room at a wooden table; my clothing was a long robed affair with long sleeves and a hood. Many what seemed to be parchment documents were beside me; my pen was a quill, and the ink container was a pot, and my light was a sole candle, and it was so very cold. I closed my eyes and said a prayer that this would go away. When I opened my eyes, I was back in my office. This episode is not one I enjoy remembering because somewhere along the line of incarnations, I lived like that and it was not a happy existence and I got the feeling in that time slip that I might not be able to "come back."

The second of Toni's very interesting reports refers to the town of Babylon in New York. Babylon is on Long Island in Suffolk County.

In 1972 I was living in Babylon, New York, in a very old farmhouse. In August that summer, I was working on a quilt and literally came to a mental block on the "finishing-stitch." The finishing-stitch is one that is the quilter's signature sign-off for good luck and blessings. I sat there and closed my eyes. When I opened my eyes, I was looking out at an open door to the prairie grass. I looked at the room I was in, and it was a sparsely furnished little house, but I got the distinct feeling I was in the West and it was late 1800s. I still had the quilt, but my clothes — I was now wearing a long gingham dress instead of my jeans and blouse. All of a sudden, a woman came up behind me and said, "Well, now, you've done that stitch so many times, it will come to you." That frightened me and I closed my eyes, opened them, and I was back in my kitchen, year 1972, but I was able to complete the stitch.

Toni's third account concerns a very mysterious train ride.

Another time, it was during the summer of 1985, I was working in New York City, which caused me to take the Long Island railroad back and forth between Long Island and New York. It was a Thursday afternoon, on the 4:11. I was reading the *New York Times*. I put the paper down and looked out of the window of the train. Something made me turn to look at the door on the train. Everyone was dressed in late 1800s clothes; the man next to me was reading a southern newspaper, and the train had turned into an ancient, coal-burning train. I distinctly felt that I was in the south on a train, and then I looked at my clothes; vintage 1800s … I did panic because I thought I was trapped in this "time warp." I took a deep breath, closed my eyes, and counted to ten. When I opened my eyes, everything was current and normal. What upset me about this incident was that I felt that I had actually really experienced that moment somewhere in time.

The particularly significant aspects of the episodes that Toni relates are the powerful feelings that the observer is not merely looking at an event in the past but is involved in it, and is at risk of being trapped in it.

One of the strangest and most persistent time slip reports centres on Wroxham Broad in Norfolk. For several centuries, sensible and reliable witnesses have reported seeing Roman soldiers and other Roman personnel either near the banks of Wroxham Broad or actually marching across dry land where the Broad should have been. The episodes are well documented. The origins of the Norfolk Broads are controversial. Academic archaeologists, historians, and geologists tend to disagree somewhat about the beginnings of the Broads, but the generally accepted hypothesis is that when the Ice Age ended, the level ground containing the Bure, Waveney, and Yare rivers was flooded because the sea level rose. As time passed, the water level went down and significant quantities of alluvium began to collect there. When the three rivers had flooded, they formed an extensive estuary: this now became a marsh, where alders and other swamp vegetation flourished before turning into peat. During the Roman occupation, the area flooded again and recreated the wide estuary. Norfolk's medieval inhabitants dug up the peat for fuel. By the end of the thirteenth century, the land was slowly subsiding and the peat digging was dying away. Sale records of the time show that £19 was paid for 400,000 blocks of peat in the early fourteenth century. Whether they were dug out as peat cuttings or for some other purposes, it seems as if the Norfolk Broads originated artificially. Is it possible that one at least was a Roman amphitheatre?

Writing in the *Archives of the Northfolk for 1603*, Benjamin Curtiss described a very strange event that he claimed to have experienced at Wroxham:

> ... in the great Broad of Wroxham, near unto Hoveton St. John. Two friends and myself were swimming across the lake from the Bure side to that opposite, when strangely enough we felt our feet touch the bottom. Now at this part there is much water, as much as twelve and in other places some fourteen feet. We kept together and presently found ourselves standing in the middle of a large arena with much seats one above the other all round us. The water was gone and we were standing there dressed as Roman Officers. What is more astonishing still, we were not surprised, neither we were incommoded by this piece of enchantment, but rather we were quite accustomed to it, so that we forget [sic] that we had been bathing. The top of the amphitheatre was all open to the sky, and many flags of divers colours floated in the wind from the top of the walls ...

The authors in Roman costumes near the Coliseum.

If we analyze what Ben Curtiss described in conjunction with what Toni experienced on the Long Island train — where she seemed to have moved through both time *and* space — it looks as if some time slips may cause spatial as well as teleological disruptions. Supposing that the Broads were only peat cuttings, and that no Roman amphitheatre was ever built in Norfolk, did Curtiss and his friends find themselves in ancient Rome rather than in ancient Wroxham?

The Wroxham Broad mystery continues with an account given by the Rev. Thomas Josiah Penston in *The Gentleman's Gazette* from April 16, 1709:

… we were holding a picnic on the banks of a beautiful lake in Norfolk about eleven miles from the ancient city of Norwich, when we were suddenly and very peremptorily ordered away by a very undesirable looking person, whose appearance and clothes belied his refinements of natural good breeding. As we were somewhat endangered by this unpleasant person's persistence, we made to go away, when suddenly we had to quickly stand aside to make passage for a long procession of regal splendour, the outstanding characters of which were a golden chariot containing a hideous looking man dressed as a Roman General, and drawn by ten white prancing stallions, about a dozen lions led in chains by stalwart Roman soldiers, a band of trumpeters making a great noise, and another band of drummers, followed by several hundreds of long-haired, partly armoured seafaring men, or sea-soldiers, all chained together.

They passed quite close to us, but no-one apparently saw us. There must have been seven or eight hundred horsemen in this long procession of archers, pike-men and ballistic machines. Whither they went or from whence they came I know not, yet they vanished at the lake side. The noise of their passing was very loud and unmistakable.

A rare and unusual poem, "Legend of the Lake," attributed to Calvert and published in the early eighteenth century, contains the lines:

While through the trees of yonder lake,
There comes a cavalcade of horsemen near.
Gaze not upon these Romans, friends,
For fear their eyes may meet with thine.
Stand back, well back, and let them pass,
These denizens of death and close thine orbs,
Lest out upon a scene of death they fall,

In hapless misery for those who play
Their parts, for nigh a thousand years.
Doomed for a term to re-enact
The life they led, the parts they played;
Go not with them, look not at them, but
Pray for them, dear friend, for they
Are dead.

Another curious reference to the Wroxham Broad time slip phenomenon is reported in the 1825 edition of *Day's Chronicles of East Anglia*. It reads: "The Royal Progress of Carausius ... has passed through ... the village of Wroxham ... on its way from Brancaster." So who was this Carausius to whom Day was referring in 1825? Marcus Aurelius Mausaeus Carausius was a member of the Menapii, a Belgic tribe occupying the northeast of France (then known as Gaul) from the first century B.C. onwards. Their territory extended from the Rhine estuary as far inland as the Ardennes, and they were also established in Ireland. Carausius made a name for himself in 286 when he fought very effectively for Maximian against rebel Bagaudae (also known as Bacaudae), who were making life difficult in Gaul. The historian C.E.V. Nixon described them as "brigands" and listed their main activities as "looting and pillaging." Following Carausius's success against the Bagaudae, he

was rewarded with command of a fleet known as the Classis Britannica, with which he suppressed the pirates and sea raiders who plied their trade between the East Anglian coast and mainland Europe. Carausius, it seemed, was a man with his eye to the main chance, and he was suspected of colluding with the pirates that he was ostensibly controlling — helping himself to their treasure, or allowing them to raid coastal villages first, then capturing them

Wroxham Broad, Norfolk, scene of repeated time slips.

after they had worthwhile valuables to hijack! For whatever reason, he rebelled against the Empire and declared himself Emperor of Britain. In view of his association with pirates, it is particularly interesting that Penston should have described "several hundreds of long-haired, partly armoured seafaring men, or sea-soldiers, all chained together ..."

Another witness to the strange Roman time slip phenomena associated with Wroxham Broad was Lord Percival Durand. According to evidence found among Durand's private correspondence from 1829, he was with a party of family and friends aboard his yacht *Amaryllis* a few hundred yards from the eastern entrance to Wroxham Broad when a curious old man appeared. Could it possibly have been the same strange old character who had appeared to Penston and his picnic party 120 years earlier? The weird old fellow who appeared to Percival Durand's yachting party on July 21, 1829, claimed to be Flavius Mantus, the *Custos Rotulorum* (from the Latin, literally "keeper of the rolls"). This was an extremely senior and important rank. Durand and his party were warned that they were trespassing on lands under the protection of the Western Emperor Carausius. According to Durand's evidence, the strange old man was suddenly transformed into a splendidly dressed Roman officer, and the waters of the broad vanished to reveal the familiar amphitheatre and imperial procession.

John and Christine Swain and their sons, from Ilminster, Somerset, United Kingdom, had a much more recent experience than the reports from Wroxham Broad. They were driving around exploring some quiet country lanes in search of a suitable picnic spot near Beaulieu Abbey, in the New Forest in Hampshire, when they passed a mysterious-looking, mist-enshrouded lake with a big boulder situated near the middle of it. Thrust into that rock — like the famous Excalibur of the King Arthur legends — was a gleaming sword.

The Swains naturally thought that it was some sort of tribute or memorial to Arthur and his knights. Fascinated by what they had seen, and remembering it vividly, they tried on subsequent visits to find it again — but they never did. Was that sword in the stone from another time or another probability track? The family had neither dreamt it nor imagined it — so where had it gone?

From Wroxham Broad and the New Forest, the trail of time slip mysteries leads on down to Pyrford in Surrey, where Mrs. Turrell-Clarke

Excalibur in the stained glass window of King Arthur's Halls, Tintagel, Cornwall.

had several inexplicable time slip experiences. Pyrford Church was built in the twelfth century, and its historic fabric has remained largely undisturbed ever since. She was going to Evensong one Sunday when the normal paved road on which she was walking changed subtly into a footpath across a field. A man in what looked like medieval peasant costume was walking towards her, but he moved aside politely to let her pass. As Mrs. Turrell-Clarke glanced down at herself, she realized that she was dressed as a nun. The experience ended as suddenly as it had begun; everything was back to twentieth-century normality, and she was dressed in her ordinary clothes again. The ancient church at Pyrford — dedicated to St. Nicholas — had enjoyed a long association with nearby Newark Priory, and only a few weeks after her strange encounter with the polite medieval peasant, Mrs. Turrell-Clarke was in church listening to the choir singing a piece of Gregorian, monastic plainsong. While they were singing, the church seemed to undergo some strange changes, much as the road had done. Lancet windows, a stone altar, and an earthen floor gave the little church a medieval look. A procession of brown-robed monks appeared singing the same music that the St. Nicholas choir were singing. After a few moments everything shifted back to twentieth-century normality again. Understandably intrigued by her strange experiences, Mrs. Turrell-Clarke began investigating the history of the village, the church, and the priory. The name *Pyrford* seems to be derived from a Saxon phrase meaning "the ford by the pear tree," and in the old folk song "The Twelve Days of Christmas" there is a reference to a "partridge in a pear tree." Some students of folklore have suggested that the pear tree is the cross from Christian tradition, and that the partridge was once believed to be a self-sacrificing bird that fed its young with its own flesh when food was scarce. The idea then has a deep religious significance.

The settlement at Pyrford seems to be very old indeed, and has a prehistoric standing stone. In addition, the circular hilltop churchyard indicates an early settlement. King Eadwig granted the manor of what was then called Pyrianforde to one of his friends in 956, and an ancient charter details the boundaries and place names that are still extant after more than a millennium.

Mrs. Turrell-Clarke's investigations revealed first that the singing monks should have been in black robes, as these were what were worn by the monks in Newark Priory. However, further detailed investigations showed that in 1293 monks from Westminster Abbey were granted use of the Priory chapel. These visitors wore *brown* robes.

When co-author Lionel was presenting his very popular *Fortean TV* show on UK Channel 4 in 1997, Colin Ayling and John England experienced a very strange time slip that was featured in one episode of the series. Colin and John were enthusiastic and experienced metal detector users, and were exploring a large field in the Shrewsbury area, which was known as Boulder Field because of the large heap of boulders piled in one corner. They found a republican silver denarius from the time of Julius Caesar, several Roman bronzes, and another denarius, this time from the period of Rutilius Flaccus, *circa* 70 B.C. Several other Roman artifacts turned up, as well as the tip of a Roman javelin into which the name "Nigel" had been punched or stamped. (The original Roman form of the name was "Nigellus," so the abbreviation is understandable.)

Suddenly, the sound of galloping horses was heard — coming straight at them. Colin and John ran for cover and heard the horses thundering over the spot where they had been working a few moments earlier. It is interesting to note that neither man actually *saw* the horses galloping past. Feeling slightly disoriented by their experience, Colin and John began making their way back to their car when they encountered what seemed to be a tightly packed, impenetrable hedge about

Co-author Lionel in the doorway of Pyrford Church, and a view of the church, where time slips apparently took place.

three metres high. They approached it cautiously, wondering if they were in the wrong field because it had not been there when they began their evening's metal detecting work. They followed this hedge, or stockade, for roughly one hundred metres before they found their way back to the car. When morning light came, Colin and John searched Boulder Field for any signs of hoofprints or the strange barrier that they had had to walk around. There were no hoofprints and no barrier fence, yet their own footprints were clearly visible, and it was easy for them to see that they had turned at ninety degrees and made a long detour around *something* that was no longer there.

The ruined priory near Pyrford, associated with reports of time slips there.

Silver denarius associated with a reported time slip.

They took their finds along to Mike Stokes, the archaeologist at the Rowley House Museum in Shrewsbury, who identified some of the artifacts that Colin and John had found as parts of Roman cavalry tackle. Is it possible that Colin and John had encountered a time slip in Boulder Field that night? Had Nigellus been one of the Roman riders galloping across time as well as turf? Is it possible that the mysterious transient barrier that Colin and John detoured around had been part of a Roman cavalry detachment's fortifications two thousand years ago?

Mrs. Anne May was a schoolteacher in Norwich. She and her husband went to Inverness to study the Bronze Age Clava Cairns, which are situated on the eastern side of the River Nairn. There are three cairns in the group: one is circular; the other two are elongated corridor, or passageway, burial sites. Most archaeologists date them between 2000 and 1500 B.C. The northeastern cairn has about a dozen standing stones around it, making a circle over thirty metres in diameter. The actual cairn itself is three metres high and between fifteen and twenty metres in diameter.

By contrast, the central cairn is circular, with access from above, and the burial chamber inside it is four metres long. The stone circle around

Mysterious and ancient: Clava Cairns in Scotland, where a time slip was reported.

this central cairn has a circumference of one hundred metres, while the cairn itself stands only a metre high and has a circumference of about fifty metres.

The third cairn is situated to the southwest; it is a corridor cairn like the first one, but slightly smaller, and its stone circle has been spoiled by a road that runs through it.

When they had finished studying the cairns, Mrs. May rested for a few moments on one of the stones and experienced something very similar to what had happened to Benjamin Curtiss, Rev. Thomas Penston, Lord Percival Durand, Mrs. Turrell-Clarke, and the two men with metal detectors in Boulder Field. Mrs. May saw a group of men from the distant past. They had long dark hair and wore rough tunics with cross-gartered trousers. As Mrs. May watched them in amazement, they dragged one of the big standing stones into its allocated place in the circle. A party of tourists then entered the site, and everything went back to twentieth-century normality.

Joan Forman, author of an excellent work on time slips entitled *The Mask of Time*, gives a fascinating and highly detailed account of a time slip that she herself experienced at Haddon Hall in Derbyshire. She saw four children playing happily together on the stone steps of the big courtyard there. The oldest was a girl who looked to be about ten years old, with shoulder-length fair hair. She wore a grey-green silk dress with a finely worked lace collar, and a white Dutch-style hat. When the experience began, Joan could see only the girl's back, but during the time that Joan was able to observe her, the girl turned so that Joan could study her face, which was quite distinctive. She had broad features, a wide jaw, and a retroussé nose. As Joan moved towards them the whole group of children simply vanished — it was as if the act of moving towards them had cut whatever telechronic* current had been making the children visible in Joan's time.

Inside the Hall, Joan looked everywhere for portraits of any of the children she had seen. At last she came across a picture of young Lady

---

* *Telechronic* is the authors' own newly minted adjective meaning "that which enables distant time to be experienced."

Grace Manners wearing the same dress, lace collar, and distinctive Dutch hat. Inquiries revealed that Lady Grace Manners had indeed been associated with Haddon Hall centuries before Joan's visit.

Haddon Hall is steeped in history. It was originally owned by William Peverel (1050–1115). His parents were supposedly Ranulph

Haddon Hall, scene of a reported time slip involving children from another century. The restrooms in Tombland, Norwich, Norfolk, where a very strange time slip reportedly took place.

Peverel and his wife, Maud, a noble Saxon lady, daughter of Ingelric, but there is a persistent romantic tradition that the nubile and beautiful Maud had an affair with William the Conqueror, who was the real father of William Peverel. For whatever reason, the Conqueror doted on William Peverel and bestowed many manors on the young man. Haddon Hall and its estate later passed through marriage to the Manners family, who became Dukes of Rutland, and this is where young Lady Grace Manners comes in.

We referred in an earlier chapter to the very strange episode in which Mrs. Turner of Lowestoft in Suffolk had uncanny knowledge of a house that she had never visited before. Her grandson, who passed that case history on to us, also had a remarkable premonition or time slip experience. This was also one of those strange, inexplicable incidents in which clothing plays a significant part. Curtiss was in Roman uniform; Mrs. Turrell-Clarke was dressed as a nun. In his very vivid dream, Mr. Turner was wearing a naval uniform and standing in line for a bus. He was asked if he was on the way to join his submarine, replied that he was, and was told by his enigmatic informant that he was too late as the sub had already left without him. Mr. Turner told us that his dream was so vivid and realistic that he became desperately anxious about missing his sub. His dream self walked the streets for hours worrying about being court-martialled and thinking about what he might be able to do to clear his name. As he wandered the street in this distraught state, he again encountered his mysterious informant, who told him that there had been a tragic accident and the sub had gone with severe casualties after a collision in the Thames Estuary. The following day, fully awake and back to his normal self again, Mr. Turner read of a submarine being lost in the Thames Estuary. He was probably referring to the HMS *Truculent,* which sank in the Thames Estuary in 1950 after colliding with a freighter.

The very strange and tragic case of the Pollock family seems to provide some evidence for reincarnation, yet there are aspects of it that are open to further careful investigation and evaluation. To all intents and purposes, the Pollock family of Hexham in Northumberland, United Kingdom, were living quiet, uneventful lives, when tragedy struck. On May 5, 1957, their two daughters, eleven-year-old Joanna and six-year-old Jacqueline, were walking to church with their young friend, Anthony Leyden. A powerful car mounted the pavement and killed all three children instantly.

Their father, John, believed in reincarnation and predicted that his wife, Florence, would soon give birth to twin girls who would be the reincarnations of the daughters they had lost. On October 4, 1958, Florence Pollock gave birth to twins — despite the fact that the doctor and gynecologist looking after her had detected only one heartbeat and one set of limbs. As far as could be ascertained there was no history of twins in either John's family or Florence's. That much of the story is certainly strange. When the twins were still under six months old, the family moved away from Hexham to live in Whitley Bay. While the twins were still not yet four years old, their parents took them back to Hexham, where they recognized both the school that Jacqueline and Joanna had attended and the house where they had lived before moving to Whitley. Another strange incident occurred a few months later. The twins were shown dolls that Joanna and Jacqueline had played with. The twins recognized and named the dolls.

Sadly, Florence died in 1979, having apparently never fully accepted her husband's ideas about reincarnation. He later remarried. His second wife and one of the twins, Gillian, share his interests in reincarnation and spiritualism. Gillian's twin sister, Jennifer, is less interested in psychic ideas.

Some investigators of the Pollock twins' "reincarnation" case would suggest that John's enthusiasm for the idea may have made the evidence less than totally objective. Had the little girls heard their parents discussing reincarnation during the time they were in Whitley Bay? Had they been "encouraged" to recognize the house and the school in Hexham? Had they been able to see the dolls or been told about them prior to "recognizing" them? It remains a controversial case, but it is nevertheless an interesting one.

One of the most interesting and intriguing time slip mysteries that has ever come our way to investigate is also one of the most frustratingly elusive and difficult to pin down. We have even broadcast on Radio Norfolk in the hope that one or more of the participants will come forward and be interviewed about the case. This mystery took place in the public restrooms in Tombland just outside Norwich Cathedral, in Norfolk, in one of the most ancient and historic parts of the city. The main participants are an elderly couple, whom we will refer to as Mr. and Mrs. Fey. There is an underground public restroom in Tombland, which, at the time of the episode, had parking spaces around it. Mr. and Mrs. Fey had finished their day's shopping in Norwich and parked on one of the spaces near the restroom so that Mr. Fey could use it. The restroom itself was a very quaint old Victorian structure, and men who use it have to descend a flight of steps — the only way in and out. There are no matching facilities for ladies on the site. Mr. Fey went down the steps, and Mrs. Fey waited in the car. Ten minutes passed and she became anxious. Had he been attacked and robbed? Had he fallen? Had he had a stroke or a heart attack? Unable to enter the men's restroom, Mrs. Fey didn't know what to do. Luckily for her, a traffic warden came up to inspect the parking time on their car (which was perfectly okay and well within its limits). She explained the problem concerning her husband and asked for his help, which was most willingly given when he realized that an elderly man might be injured or ill and needing medical attention. He searched the three washroom cubicles — all empty. There were six porcelain places in the restroom, and three wash basins along one wall. It was a simple enough place to search. The traffic warden found no sign of Mr. Fey, no signs of a struggle, no blood anywhere to suggest that someone had been attacked. Mrs. Fey was pleased in one way to hear his report: at least she knew that her husband was not lying ill or injured down there. As far as she was aware she had stayed watching the top of the stairs for his return. Had she glanced away at any time? She didn't think so. But was it possible that he had come up while she was looking elsewhere and had gone to get something that they had forgotten during their shopping trip? Could she have missed seeing him? If so, where was he now? In one of the shops across the road? It didn't seem likely. She was deep in thought when he appeared at the head of the stairs, walked rather unsteadily to their car, opened the door, and flopped into the driving seat.

"I'll explain in a minute, when I've pulled myself together," he said quietly. Mrs. Fey took his hand comfortingly and waited. "I went down and used the restroom, washed my hands, and came up again," he began. "When I got to the top, you weren't here, the car wasn't here; everything looked strangely different, and there were cars of a type I have never seen before gliding past on that road without making a sound!" He looked very stressed and bewildered. "I just didn't know what to do. The only thing that hadn't changed was the steel handrail that I was still holding, and the steps back down behind me. I thought hard about what was the best thing to do, and then decided to go back down again. I opened one of the cubicles, lowered the seat cover, and sat there to rest quietly and try to puzzle out what had happened. Finally, I decided to come back up and see if things were still weird, or whether they were normal again. And here you were!"

Mrs. Fey told him about how worried she had been, and how a helpful traffic warden had gone down to look for him — only to find the place completely empty. It can only be assumed (always presuming that the story can be verified) that when the traffic warden went down to look for him, Mr. Fey had already come up into another probability track in which, in some future city, cars of an unknown design were gliding silently past him.

Silhouette of Bishop Bathurst, father of Benjamin, the missing diplomat. The authors are very grateful to Mr. Neal Wood for permission to reproduce this picture.

The strange case of Mr. and Mrs. Fey raises the query of how many inexplicable appearances and disappearances could be the results of time slips, and this raises the Fortean mystery of the vanishing diplomat: Benjamin Bathurst, son of the then Bishop of Norwich.

An interesting factor of the Bathurst case is that it falls into the category of the "now-you-see-it-now-you-don't" mystery. The famous Borley Rectory hauntings serve to illustrate the main features of this type of ongoing investigation. Psychic investigator Harry Price first visited the Rectory in 1929, took up a tenancy there from 1937–1938, and wrote several books on its Victorian history and his own later investigations. Judged by any objective criteria, and with due respect for the statements of various witnesses over a long period, it certainly seems that there were paranormal and anomalous phenomena connected with Borley Rectory long before Harry Price began his investigations. He was later accused of faking some phenomena, and in one disreputable episode he seems to have been caught with his coat pocket full of pebbles, which he was flicking surreptitiously to create "poltergeist" phenomena! However, long after the Rectory burnt down, and well after Price's death in 1948, other investigators have reported strange phenomena in the area —

Borley Rectory — once described as the most haunted house in England.

particularly from the church. The pattern consists of interesting psychic phenomena, followed by accusations of fakery, followed by more (apparently genuine) psychic phenomena.

Bathurst's disappearance at Perleberg on the night of November 25, 1809, certainly follows this Borley-type pattern. Charles Fort recorded Bathurst's sudden disappearance as happening when Bathurst "walked around the horses and was never seen again." Other investigators have looked deeply into some of the details associated with the mystery of Bathurst's disappearance and have come up with interesting queries and problems. Sabine Baring-Gould wrote two fascinating books entitled *Historical Oddities*, which were published in London in 1889. His first volume included the mystery of Bathurst's disappearance, which he had published earlier in volume 55 of the *Cornhill Magazine*. Baring-Gould did much of his research via the earlier work of Mrs. Tryphena Thistlethwayte, the missing man's sister, who gave a full account of Benjamin's disappearance when she wrote up her father's memoirs in 1853.

Borley Church. After the Rectory burned down, strange phenomena were reported from here.

Benjamin was born in 1784, third son of Dr. Henry Bathurst, Bishop of Norwich. Benjamin married Phillida Call, daughter of Sir John Call, a wealthy Cornish landowner. It is relevant to Benjamin's disappearance that he was wealthy, influential, and very well connected. His disappearance — especially as he was a member of the Diplomatic Corps — set powerful forces in motion to find him and, if humanly possible, bring him safely home again. Earl Bathurst, a distant relative of Benjamin's, was Secretary of State for the British government's Foreign Affairs Department in 1809 and sent Benjamin on a mission to Vienna. The idea behind it was to try to persuade the Austrians, under Emperor Francis I (1768–1835), to attack the French. Austria agreed and duly sent troops into Italy. They were soundly defeated by Napoleon at the Battle of Wagram, and Austria sued for peace. Bathurst was now in an untenable position in Napoleon's Europe, and he became convinced — not without reason — that the French and their allies wanted him dead. His first thought was to head south and get back to the United Kingdom via the Adriatic. That might well have been a better idea for him, but he discarded it in favour of the alternative route through the notionally neutral Confederation of German States, via Berlin and Hamburg.

The whole tragedy took on the air of a James Bond story when Benjamin pretended to be a German merchant named Koch. Fatefully, he reached Perleberg around lunchtime on November 25, 1809. French Napoleonic troops were just a few miles away at Magdeburg and Lenzen. The town of Perleberg was not a safe place: it was full of ex-soldiers, deserters, and minor and major criminals of various kinds, all of them pretty desperate for money. In the Perleberg of 1809, life was relatively cheap. Bathurst and his two servants had dinner at the White Swan Inn. The diplomat spent some time alone in his room, writing some papers and carefully burning others. From here on, the narrative splits into several versions. In one, the best known and most mysterious, Bathurst walked around the horses and vanished as if he had never existed. Another account has him in the kitchen with a group of suspicious characters, in front of whom he foolishly takes out his purse (which had a significant sum of money in it) along with an expensive watch. This version suggests that he was dragged out by some of the criminals, robbed, and murdered. What subsequently became of his body is not made clear.

In another version, Bathurst seems to have absented himself for an hour or so, while his servants waited patiently for him. At his request, Captain Klitzing, a local military officer, had posted two guards to look after the young diplomat, but they had gone by 7:00 p.m. and Bathurst's disappearance was two hours after that. Bathurst himself seemed to be in a very worried state prior to his disappearance, and there is evidence that he believed a sinister French double agent, known as the Count d'Entraigues, was after him.

Young Phillida Bathurst bravely went in search of her husband, and got as far as an audience with Napoleon. He assured her that whatever had happened to Bathurst, it was none of his doing. To give Captain Klitzing his due, he made an extensive and prolonged search for the missing diplomat in and around his Perleberg jurisdiction. This search included dragging the River Stepnitz. Then suspicious objects began to turn up. Bathurst's expensive coat was found. Later his trousers were discovered — complete with bullet holes, but no corresponding bloodstains.

Over the ensuing years a hidden skeleton or two turned up in the Perleberg area, but nothing that could definitely be identified as Bathurst's remains. The mystery was never satisfactorily solved, and the controversy continues as to whether the missing diplomat vanished suddenly and mysteriously — perhaps the victim of a time slip — or whether he was abducted, robbed, or murdered.

The sudden and inexplicable arrival of young Kaspar Hauser in Nuremberg on May 26, 1828, created as great and persistent a mystery as the vanishing of Benjamin Bathurst had done twenty years before. A kind-hearted shoemaker named Weissman took the boy to see Captain Wessenig, as the bewildered youngster was carrying letters addressed to the captain of the 4th Squadron of Dragoons in Nuremberg. These letters gave his date of birth as April 30, 1812. Kaspar seemed to be badly traumatized, and to suffer from some form of mental illness. He was unable to speak coherently when he first arrived in the town, but gradually unfolded a very curious story. He had, he said, been kept in a small, dark hole with straw as its only furnishing. From time to time he had been drugged, and when he recovered consciousness, he had been given clean clothes. It was during these periods of unconsciousness that his hair was cut.

Kaspar was transferred to the care of a kindly schoolteacher named Friedrich Daumer, in whose hands the strange boy made excellent progress.

There were rumours that he was the rightful heir to the Grand Duke of Baden, Karl Friedrich, by Stéphanie de Beauharnais, adopted daughter of Emperor Napoleon, and that he had been smuggled away so that Leopold I could succeed Friedrich instead. The finger of suspicion was pointed at Leopold's mother — but nothing could be proved at the time.

A hooded and masked intruder apparently attempted to kill Kaspar on October 17, 1829, but he was only wounded and recovered well. A successful attempt on his life was carried out on December 14, 1833. After being lured to the Hofgarten in Ansbach, Kaspar claimed that he had been stabbed by a masked man, and he died three days later. Those who felt that Kaspar was only seeking attention claimed that he had probably wounded himself in 1829, and had stabbed himself to create more interest in 1833. However, there are reasons to suspect otherwise. His headstone reads: "Here lies Kaspar Hauser, riddle of his age: his birth unknown, his death a mystery."

Twenty-first-century scientific evidence from the Forensic Medicine Department of the University of Münster, based on careful testing of several samples of Kaspar's hair, suggested that he *was* related to Stéphanie de Beauharnais and could have been the hereditary Prince of Baden. How did the unfortunate boy get into what he himself described as a straw-filled hole for so many years? Is it even remotely possible that Kaspar was yet another hapless time slip participant, rather than the victim of cruel dynastic intrigue? Did he appear suddenly in Nuremberg that morning because whoever had kept him for so many years in the straw-lined hole had brought him there secretly in 1828? Or was a dark, straw-filled hole the nearest description that Kaspar's traumatized and hopelessly confused young mind could find to describe some weird limbo resulting from his involvement in a time warp? An unexplained *appearance* is just as great an enigma as an unexplained *disappearance*. Maybe Kaspar and Benjamin Bathurst have more in common than has yet been recognized.

Kaspar Hauser. Did he travel through time to reach Nuremberg?

Disappearances are even more significant when more than one person is involved. The unsolved mystery of the vanishing complement of the *Mary Celeste* involved a total of eleven people: Captain Benjamin Spooner Briggs, aged thirty-seven; Mrs. Sarah Elizabeth Briggs, the captain's thirty-year-old wife; their two-year-old daughter, Sophia Matilda; three Dutch sailors; the American steward and cook; First Mate Albert Richardson; one German sailor; and the Danish second mate. The *Mary Celeste* had been built in Nova Scotia in 1861, and had been launched from there as the *Amazon*. She was 103 feet long with a tonnage of 282. The sturdy little brigantine was renamed *Mary Celeste* in 1869 after a series of disasters and misfortunes, including the death of a previous skipper. The *Mary Celeste* set sail from New York on November 7, 1872, bound for Genoa with a cargo of industrial alcohol, which could be volatile and dangerous to handle. On December 4, 1872, Captain David Reed Morehouse of the *Dei Gratia* sighted the *Mary Celeste* drifting sporadically as though no one was at the wheel. Morehouse was a friend of Briggs, and realized at once that something was wrong. The success of the attempted rescue hinged on the great physical strength, courage, determination and tenacity of one of the best sailors afloat: Captain Morehouse's first mate, Oliver Deveau. There was nothing Deveau wouldn't tackle, and, once committed, he'd see it through. Almost anything could have awaited the would-be rescuers, when the fearless Deveau and another seaman from the *Dei Gratia* rowed across to investigate the mystery of the *Mary Celeste:* plague, mutiny, murder, pirates, or slave traders (the attractive Sarah Briggs would have fetched a good price on the Barbary Coast).

The powerful and intrepid Deveau pumped out a few feet of water from the hold and examined the *Mary Celeste* from stem to stern. Everything seemed safe, sound, and seaworthy. There were ample supplies of food and water on board. There did not seem any compelling reason why the ship should have been abandoned.

Various attempts to explain the mystery over the years have included the idea that alcohol fumes leaked from some of the 1,700 barrels in the hold and blew the hatch covers off, leading to the fear that the entire cargo was going up at any minute. With the safety of his much-loved wife and baby daughter to consider, did Captain Briggs

The brigantine *Mary Celeste*.

Captain Benjamin Spooner Briggs, one of those who vanished from the Mary Celeste in 1872. Did they fall through a chasm in time?

Captain Briggs's first mate, Alfred Richardson.

Captain David Reed Morehouse of the Dei Gratia, who came to help the *Mary Celeste*.

First Mate Oliver Deveau of the Dei Gratia — a powerful and fearless mariner who played a major role in salvaging the *Mary Celeste*.

give the order to abandon ship? The small, inadequate lifeboat was missing from the deck of the *Mary Celeste*. If there was a long line attaching that lifeboat to the stern of the brigantine, and a sudden powerful gust snapped it, what chance of survival would the complement of the *Mary Celeste* have in heavy seas?

The brig sailed on for another eleven or twelve years, finally being run up deliberately on to the Rochelois Reef in Haiti, in the hope of claiming insurance. Clive Cussler, representing the National Underwater and Marine Agency, found the remains of the *Mary Celeste* in 2001.

Is it remotely possible that those eleven souls were the victims of a time warp of some type? If they were, where did it take them?

The next mysterious disappearance that could, perhaps, have been the result of a time slip concerns the legend of the fiddler of Binham Priory, in Norfolk. Not far from the village of Binham stands an early Bronze Age burial mound, known since medieval times as Fiddler's Hill. Centuries ago, the villagers had been greatly disturbed by the frightening apparition of a phantom monk or friar, who was referred to in the legend as the Black Monk. Witnesses claimed that they had seen him gliding out of the entrance to a tunnel, which traditionally led from Binham

Priory to the Shrine of Our Lady of Walsingham, barely five kilometres away. This Walsingham Shrine has an intriguing and mysterious history. In 1061, the Lady Richeldis, a very pious widow with a young son, prayed that she would be shown some special way to honour and pay tribute to the Virgin Mary. In response to Richeldis's fervent prayers, Mary appeared to her and transported her in spirit to Nazareth, where she showed Richeldis the house in which the Archangel Gabriel had appeared to her and proclaimed the future birth of Christ. Mary then requested Richeldis to build a shrine at Walsingham with exactly the same dimensions as the house in Nazareth where the Annunciation had taken place over a millennium before. This was duly completed, although one version of the legend says that when the builders were experiencing problems, Richeldis prayed again and angels appeared to finish the work. There is often a core of historical truth in legends and folklore. Richeldis's spiritual journeys to a house in Nazareth that existed a thousand years before her own eleventh century could, perhaps, have been the result of time slips that took her through both time and space.

Mrs. Sarah Briggs, Captain Briggs's wife.

Binham Priory in Norfolk, where a travelling fiddler vanished in an ancient and mysterious tunnel. Was it a tunnel in time?

Other legendary miracles associated with the Walsingham Shrine include the escape of King Edward I from serious injury or death, when falling masonry narrowly missed him.

The Binham tunnel, then, can be seen to have close connections with the allegedly haunted Priory, the Bronze Age burial mound, and the Shrine of Walsingham with its reputation for miracles. An itinerant fiddler, accompanied by his faithful dog, visited Binham and was intrigued by the villagers' accounts of the spectral monk who emerged from the mysterious tunnel. The fiddler agreed to explore the tunnel, playing as he went, so that the villagers could hear the sound of his violin. The sound was plainly audible to them as they followed him on the surface for several hundred metres; then it suddenly stopped. No one was willing to enter the tunnel in an attempt to rescue the fiddler, but some hours later his dog emerged looking bedraggled and terrified. Was the intrepid fiddler the victim of a time warp?

Madonna of Walsingham, where the Binham tunnel was said to lead.

Just as various skeletons were discovered in Perleberg, any one of which might have been Benjamin Bathurst's, so, centuries later in 1933, during a road widening scheme, skeletons were found in what might have been part of the ancient tunnel from Binham to Walsingham. Could one of those skeletons have been all that was left of the intrepid fiddler — or had he really vanished through a time warp?

Blakeney in Norfolk is also reputed to have its fair share of mysterious tunnels, similar to the one connecting Binham and Walsingham. The Blakeney passageways connect the Hall in the nearby village of Wiveton with the old Carmelite Friary and the Guildhall in Blakeney. The Guildhall was once used as mortuary for the bodies of drowned sailors. Just as in the Binham legend, a dauntless fiddler and his dog went to explore the tunnel and the fiddler was never seen again. Evidence for the Blakeney tunnels is quite significant. In 1921, when the historic Crown and Anchor inn was demolished, tunnels were found underneath it. Also back in the 1920s, workmen digging on Mariners Hill got down three or four metres and encountered the barrel-arched roof of an old tunnel. The pressure on them to complete their task was such that they were not able to explore it, so it was just covered over again. Another mysterious tunnel was found in Little Lane in Blakeney, which seems to have connected the houses of two of the main nineteenth-century ship owners who lived opposite each other. Further tunnels were found during work being carried out in the village in the 1970s.

Three hundred miles to the west of the Norfolk tunnels, similar mysteries are associated with the strange old hermitage that is cut into the rock at Bridgnorth, Shropshire, United Kingdom, about a mile from the town itself, which lies in the valley of the River Severn. A prince of the old Kingdom of Mercia, Aethelward, a grandson of Alfred the Great, lived there in the tenth century, and not far from his hermitage there are two so-called witches' caves. Legend has it that there are mysterious staircases in these caves that psychic visitors can see, but others can't. The legends also tell of enigmatic tunnels running back from various caves in the area and connecting with one another deep below the ground — not unlike the Cretan Labyrinth. Just as at Binham and Blakeney in Norfolk, there are legends here in the Bridgnorth area of fiddlers vanishing in these strange tunnels. Just as water featured prominently in the Wroxham Broad time slip, and at the events in the coastal villages of Binham and Blakeney, so water appears again in the River Severn that runs close to Bridgnorth.

The mysterious Green Children of Woolpit in Suffolk are yet another possible time slip phenomenon with strong water connections. During the reign of King Stephen (1135–54), according to the chroniclers Ralph of Coggeshall and William of Newburgh, two green children — a boy and a girl — appeared in the harvest fields at Woolpit.

The village name came either from "wolf-pit" or "wool-pit," and there is a small lake in the village that we photographed when doing our research there. Both Ralph and William are credible early historians, and the basic facts as they have recounted them seem to be that the boy and girl appeared very suddenly and mysteriously. The children were uncertain about where they had come from, and referred to it as St. Martin's Land. They had difficulty with language at first — just as

The sinister pit that gave Woolpit in Suffolk its name. Did these strange waters have a mysterious association with a door through time?

The town sign of Woolpit, showing the riddle of the two green children. Did they step out of time to reach Woolpit?

Kaspar Hauser had done — but eventually settled into normal twelfth-century English village life. The girl eventually lost her green colouring and married a man from King's Lynn. The boy, however, sickened and died. The unanswered question of their mysterious origins remains. Were they also the victims of a time slip? Or were they merely unfortunate children who had been abducted and forced to work in a medieval copper mine (which could have accounted fro their green colouring)? Were they refugee Flemish children whose parents had been murdered, and who had wandered, badly undernourished and consequently suffering from green sickness? This is a form of chlorosis (or chloremia) characterized by a lack of hemoglobin in the red blood cells: children and teenagers are more prone to it than adults — especially if they are subsisting on an inadequate diet, as the Woolpit children may have been.

While working on their *Fortean TV* series at various sites in and around the London area, the authors were being driven to their next location by a studio driver who shared their interest in all things paranormal and anomalous. Knowing that co-author Lionel was an enthusiastic biker, the driver recounted an episode in which he had been driving towards Maidenhead in a Standard Vanguard: a big, powerful limousine that was very much sought after in the 1950s and '60s. Maidenhead, in the county of Berkshire, is not far from ancient Royal Windsor, but dates back only to the thirteenth century, which is relatively young by the standards of English settlements. The modern town takes its name from a timber wharf — called a hythe in those days — that ran along the bank of the Thames adjacent to the bridge that had been built there during the first part of the thirteenth century. As this hythe was new, it was referred to as the maiden-hythe, from which the name Maidenhead eventually developed. Our driver recalled that he and a companion were driving their Vanguard towards Maidenhead on a dark, foggy night, when without warning a fast-moving motorbike came straight at them. The driver braked hard, but the bike hit them on the driver's-side front wing. Shaken, but determined to do everything possible to help the biker — whom they felt sure must be badly injured or dead — the driver and his friend searched everywhere for man and machine. They found no sign of either, but the front wing of the very sturdy Standard Vanguard was badly dented. They drove on into Maidenhead and reported the accident at the first police station they saw in the town, requesting an officer to accompany them back to the scene of the accident to have another look for the motorcyclist. The sergeant at the desk was a relatively young officer; he was taking down all the details when an older constable who had served in the area for twenty years or more said, "I'll go if you like, Sarge, but there won't be anyone there! A motorcyclist *was* killed there many years ago, and these gentlemen are the fifth or sixth honest and concerned witnesses who have come in here to report that same accident over the years I've been at this station."

How is the Maidenhead motorcycling tragedy to be explained? Was it another time slip, and, if so, who found himself in the wrong place at the wrong time? Was it the unfortunate motorcyclist who found himself on a road that belonged in another time — and is that why he was unable to avoid the car that killed him? Is it reasonable to assume that in his own time — say the 1930s or '40s — he had experienced a time slip and crashed into a vehicle from another time or another probability track? If the shocked driver of the car that had collided with him had failed to find him, was it because the time slip had corrected itself and the motorcyclist's body lay in his own time, while the motorists were looking for him in their time? Did the dead motorcyclist look to the police officers and ambulance crew who found his body as if he must have been the victim of a hit and run driver who had failed to report the accident? It would be very, very difficult — if not impossible — to find a single case in the archives of anomalous phenomena in which a ghost, apparition, spectre, or phantom had brought about a significant physical change such as the dent in the wing of the Standard Vanguard. But a time slip that brought the bike and car together with fatal consequences for the biker could well inflict real physical damage on the car as well as the bike. The fact that the constable could remember being called out several times and finding nothing seems to suggest that there was something very strange in the area that was conducive to the warping or slippage of time in that location. It should also be remembered that a significant quantity of water was again close to the scene: a thirteenth-century bridge crossed the River Thames next to the hythe that had given Maidenhead its name.

Chapter 9

MORE CASE STUDIES OF THE TRICKS TIME PLAYS

WHILE THE AUTHORS were carrying out research in the cata-combs of Rome in the summer of 2006, working in the Catacombs of St. Callixtus, also known as the Catacombe di San Callisto, co-author Lionel (who is generally reckoned to be tougher than reinforced con-crete) had an inexplicable paranormal experience.

Pope Callixtus served from 217 until 222, but prior to that he had had a somewhat checkered career. When he was a young slave, his mas-ter, Carpophorus, had rather unwisely put him in charge of a bank. Callixtus lost a great deal of the depositors' money and fled. He was recaptured, but leapt overboard near Portus. Rescued from the sea, he was taken back to Carpophorus, who then listened to the creditors' requests and released Callixtus again in the hope that he might be able

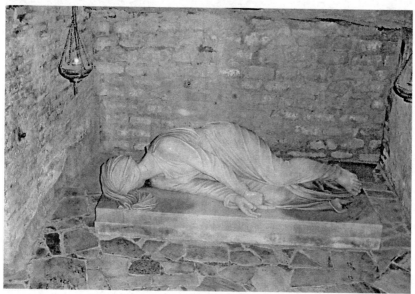

Statue of Saint Cecilia in the Catacombs of St. Callixtus in Rome.

to recover some of the money he'd lost. He promptly got into a fight in a synagogue and was arrested again. He was later sentenced to work as a slave in the mines of Sardinia — tantamount to a slow death penalty. Emperor Commodus had a beautiful mistress named Marcia, whose kindness and humanity matched her sensuous loveliness. She begged Commodus to release the Christians in the mine, and he agreed. Callixtus was then in very poor health, but by the generosity of other Christians — including Pope Victor I, who granted him a pension — he was sent to Antium to recover. Callixtus later served as a deacon, and Pope Zephyrinus (199–217) put him in charge of the burial chambers situated along the Appian Way.

These amazing catacombs held the remains of nine popes in the third century, but they slid quietly out of both ecclesiastical and secular history until the nineteenth century when Giovanni Battista de Rossi, the renowned archaeologist, found them again.

Callixtus succeeded Zephyrinus as Pope and displayed a very commendable, merciful and liberal attitude to all repentant sinners. The more rigorous Hippolytus and Tertullian bitterly criticized Callixtus for this,

Three figures on the wall of the Catacombs of St. Callixtus in Rome.

and were particularly upset when he admitted penitent adulterers and fornicators to Mass. Sadly, despite (or perhaps because of) his enlightened liberal views, Callixtus was martyred in 222; according to some traditions, he was thrown down a well. His Feast Day is celebrated on October 14.

While working in the catacomb that bears Callixtus's name, Lionel was behind the rest of the party, and almost completely separated from them. Patricia was strategically stationed at a ninety-degree corner of the catacomb where she was still just visible to Lionel, and from where — by looking ahead of her and to the left — she could also see the rearmost members of the advance party. The authors are convinced beyond a shadow of doubt that no one in the party was behind Lionel in the catacomb. Patricia was a good thirty metres ahead of him, and the advance party were another thirty metres ahead of her. Lionel then became acutely aware that someone very tall was standing silently behind him and looking over his shoulder. Thirty years and more of visiting and investigating some of the strangest sites on Earth tend to give a man more than his fair share of bold self-confidence — and if whoever (or whatever) was behind him had been merely a prospective mugger or an opportunist thief, he had unquestionably picked the wrong target. (Lionel was a wrestler before becoming a third Dan martial arts instructor.)

The tall stranger behind him in the eerie darkness of the Callixtus Catacombs was not of this Earth, but he was nothing hostile or negative. If he gave off any psychic atmosphere at all, it seemed to be *curiosity.* He seemed to be asking politely enough who Lionel was and what he was doing there. Lionel also got the impression that the entity was probably an ordained deacon or priest. He was wearing a tall, pointed hat — like a traditional wizard from legend and folklore — and a long cape, which, together with the hat, gave the outline of a tall upright cone. The cloak and hat were black, but they shone, gleaming and glistening as though something bright and sparkling was woven into them. When Lionel turned to look more closely at the entity, he could see nothing. It was one of those apparitions that is restricted to peripheral vision. Because, as a priest, Lionel is frequently called upon to conduct funerals, and regards comforting and helping the bereaved as one of the most important parts of his priestly work, he wondered whether the entity that had looked over his shoulder, down there in the solemn silence of the Callixtus

Catacombs, had also been a priest — one who had laid to rest the mortal remains of those that lay there. Thinking through the experience in retrospect, and trying to understand what had really happened there in the silent darkness of the catacomb, Lionel is becoming increasingly convinced that it was an early Christian funeral priest whom he met.

This, then, raises the question not of an encounter with a ghost in the traditionally accepted sense of a spirit or phantom, but of a time slip. Did a priest from the third or fourth century encounter a fellow priest from the twenty-first century? Did a man who had done his best to help the bereaved seventeen centuries ago glide through a mysterious portal in time to encounter a kindred spirit doing that same work today? Whatever the final explanation of their encounter may turn out to be, the best kind of evidence for investigators of the anomalous is this first-hand and personal kind.

The story didn't quite end there. For the next thirty-six hours or so, even after the authors had arrived back at their hotel, Lionel continued to catch peripheral glimpses of the entity from the St. Callixtus Catacomb — and he wasn't alone. The peripheral glimpses of something that wasn't physically there were lower down in many of these later appearances. It was almost as if the funeral priest from the catacomb had — like St. Francis of Assisi centuries later — been a lover of animals. Some of the later peripheral glimpses that Lionel experienced after the catacomb phenomenon were knee high — like sheepdogs or tamed, docile wolves. Within three or four days of our catacomb research, even these smaller peripheral visions had faded and gone. Had they too stepped through time with their mysterious master and returned with him to the third or fourth century? Or was it the other way around? Had co-author Lionel passed briefly into a time seventeen centuries behind us? Was it his unwitting appearance in that other priest's time that had created the mysterious entity's feeling of curiosity? The whole episode makes more sense as a time slip than as an encounter with a ghost. What was the mysterious common denominator that had created a strange bond of mutual attraction between them? As a laid-back, liberal, modernist, academic twenty-first century theologian, Lionel has a great deal of sympathy for Callixtus — but not too much for the grimly pious, dogmatic Tertullian and

Hippolytus. Could that strange figure looking over Lionel's shoulder in the catacomb have been Callixtus himself, from the days when he was the deacon responsible for it? Did Callixtus sense that this British priest — visiting these ancient Roman burial places from a century seventeen hundred years ahead of his own — was a tolerant kindred spirit, potentially a theological comrade-in-arms who would stand beside him in his dispute with bitter opponents who held cruder, narrower, less merciful views?

Whatever happened in that catacomb differed greatly from the mysteries surrounding reports of reincarnation, and the riddle of reincarnation is itself of a different order from the enigma of possession — yet there is a curious nexus between the two categories.

Cases of apparent reincarnation can clearly be related to the unsolved mysteries of the nature of time. If in the twenty-first century we "recall" being a Neanderthal hunter, an Egyptian priest, a Roman slave trader, a Viking marauder, or a bowman from Gwent at Agincourt, then time is the subway along which our immortal spirit travelled from that past life into our present existence as a checkout operator in a supermarket or a postal worker. But what of a case like that of Lurancy Vennum and Mary Roff of Milford and Watseka in Iroquois County, Illinois, United States, in the second half of the nineteenth century? Lurancy was a happy, healthy, normal teenager until July 1877. She suddenly announced that mysterious psychic voices were calling her name ... then she began having inexplicable seizures during which she saw visions of heaven, angels, and people whom she knew to be dead but who were now very much alive again. These strange visions were very much like the mysterious experiences that affected St. Juliana of Norwich, also known as Julian. Juliana was a mystical saint who lived in an anchorage (a type of hermitage) very close to the ancient church of St. Julian in Norwich, Norfolk. She was born in 1342, and there is evidence that she was still alive in 1413. Her deeply religious book *Sixteen Revelations of Divine Love* was the result of revelations that she believed came to her from God in 1373 when she was thirty-one years old. During one of her ecstatic visions, Juliana had a vision of Heaven, and when asked to describe what the life to come was like, she answered, "All shall be well; all shall be well; and all manner of thing [sic] shall be well."

When young Lurancy Vennum saw heaven and the angels, she seems to have had an experience resembling Juliana's. As the seizures got worse and became more frequent, some friends of the Vennum family, Mr. and Mrs. Asa Roff from Watseka — six or seven miles from Milford — suggested calling on the services of Dr. E.W. Stevens, who came from Janesville in Wisconsin.

It is important when investigating the case of Lurancy Vennum to consider the predominance and popularity of spiritualism in the United States during the second half of the nineteenth century. It would not be an exaggeration to suggest that the majority of Americans at that time almost *expected* to hear about incidents involving ghosts, phantoms, apparitions, poltergeists, possession, reincarnation, and a score of other paranormal manifestations. It is, of course, perfectly possible that all that psychic expectation actually played a part in *creating* the manifestations that so many people were expecting to see and hear. Alexandra David-Neel and her Tibetan tulpa — the jolly, smiling monk whom she created, but who became increasingly sinister and uncontrollable and was very hard to get rid of — may provide a relevant parallel here. Another interesting parallel exists in the case of the totally imaginary Philip of Diddington Hall. Members of the Toronto Society of Psychical Research, and their eminent scientific adviser, Dr. A.R.G. Owen of the University of Toronto, conducted a fascinating experiment within which the life story of the non-existent seventeenth-century character Philip of Diddington was central. Philip's tragedy was to be married to a frigid wife, Dorothea, and to have found a beautiful gypsy girl, Margo, whom he secretly established in the gatehouse as his mistress. Dorothea found her, accused her of witchcraft, and had her burnt at the stake. The gutless Philip did nothing to save her, and later, overcome with remorse, he flung himself from the battlements of Diddington Hall (a far cry from Lancelot's noble death-or-glory response when, in the Arthurian saga, he risked life and limb to rescue his beloved Guinevere from a similar fate).

The Toronto experimenters held a series of séance-type meetings in which they tried to contact the "ghost" of the fictitious Philip — who eventually turned up! But what *was* it that caused the knocks, the violent table movements, and the other strange manifestations that the Toronto experimenters recorded? Was the so-called Philip evidence of the creative and dynamic power of their group subconscious? Was he a tulpa-like creature — a much more physically powerful version of Alexandra's unpleasant little monk? Or, as some investigators have suggested, was there another powerful psychic entity there that had latched on to the Philip narrative ? Philip may have been no more than the lines in a Shakespearean folio describing Macbeth's character and murderous activities, yet a brilliant professional actor can bring Macbeth to life so vividly that the audience — at least for the duration of the performance — feels that he is a real person. Are there psychic actors as well as physical ones? Was someone — or some*thing* — playing the role of Philip of Diddington, which the Toronto research team had written so convincingly?

For whatever reason, there were some remarkable and persistent psychic manifestations surrounding Lurancy Vennum. On January 31, 1878, Dr. Stevens duly called on Lurancy and her family, and brought Asa Roff along to make the introductions. When they arrived, Lurancy seemed to be possessed by more than one psychic entity, including a man named Will Canning and a woman called Katrina Hogan. Dr. Stevens hypnotized her, whereupon Lurancy maintained that she had been possessed by evil spirits, malevolent sprites, or demons. Then her mood changed, and she said that there was one other spirit that wanted to use her body, to possess her and control her. Stevens asked who this was, and Lurancy replied, "Mary Roff."

"Mary was my daughter," said Asa Roff in amazement. "She died before her nineteenth birthday — and that was twelve years ago." Mary's tragic and untimely death had taken place in the summer of 1865, when Lurancy was already a year old, so this was not a classic case of reincarnation. Mary had also suffered from trance-like seizures — like Juliana of Norwich, and Lurancy. In fact, it was because of those seizures that Mary had enjoyed something of a reputation as a person with psychic gifts. After various family discussions it was felt that as Mary had been a good person, and had suffered considerably during her short earthly life because of the seizures, her spirit would be ideal therapy for the troubled Lurancy.

During the next twenty-four hours it was as if the spirit of the dead Mary had completely taken over Lurancy's body, while Lurancy, as a personality, had apparently ceased to exist. Lurancy as Mary was quiet, gentle, polite, and self-effacing, but she kept begging the Vennums to allow her to go home. A few days after the strange personality transition had taken place, Mrs. Roff and her married daughter, the late Mary's sister, Nervie, came to visit the Vennums. As soon as Lurancy saw them she called out, "Here's ma and my sister, Nervie." More urgently than ever, she again begged the Vennums to let her go home with her family. Shortly afterwards, they agreed, and Lurancy — still convinced that she was Mary — went to live with the Roffs. She stayed with them for several months, remembering and recognizing Mary's old friends and things that Mary had made and used. When the Vennums called to check that Lurancy was all right, she did not recognize them as her parents, but slowly came to accept them as new friends and treated them accordingly.

On May 7, 1878, Mary told the Roffs that she was going away, and that Lurancy was coming back. Then she simply sat down and closed her eyes. When she reopened them, she looked bewilderedly at Mrs. Roff and said, "Where am I?" To all intents and purposes, she was Lurancy once more. For the next few weeks, the personalities of Mary and Lurancy alternated, but Lurancy gradually became dominant. The Mary personality almost disappeared, but not quite. Occasionally, over the years, it would seem as if Lurancy was in the kind of trance that some psychic mediums enter during séances, and then Mary would come through again.

The remainder of Lurancy's life was happy and normal: she married farmer George Binning in 1882 when she was seventeen. They later moved to Rollins County in Kansas and raised a family together.

Working on the assumption that the facts in the case of Mary Roff and Lurancy Vennum were more or less as reported by Dr. Stevens and the families involved, what might they *mean* in terms of time theory? Straightforward reincarnation — in which body A has died before the spirit that formerly lived in it has moved into newly conceived, or newly born, body B — involves a theory in which time can be represented as one of the possible media in which disembodied souls or spirits exist. Does this necessarily involve a theory of time as a unidirectional arrow-flow? Suppose that immortal spirit Q occupied physical body A from 1810 until 1897, when physical body A died and decayed back to its elements. It may then be supposed that Q waited in the time medium, or "stepped outside time," until 1903, when it re-entered the newly born body of baby B and lived in it until 1979. The occupancy of body A, the death of body A and the subsequent entry into body B are all sequential. But in the Vennum and Roff cases, Mary Roff died approximately a year *after* Lurancy was born.

Consider the possibility that time warps, time distortions, and other teleological faults and irregularities can and do occur periodically. What if arrow-line, unidirectional time can experience convoluted retractions? Imagine a very flexible elastic cylinder. A shock or trauma in time could cause units 34, 35, and 36 of that cylinder to retract and slide up inside units 31, 32, and 33. Time unit 31 now coexists with time unit 34; 32 is with 35; and 33 is with 36. If an eternal spirit such as that which had occupied the body of Mary Roff for over eighteen years is propelled back into time when that body dies, it may legitimately enter the newly conceived body of Lurancy Vennum. Because of the retracted, convoluted time irregularity, however, another immortal spirit is already inside that body. If time had not been behaving anomalously just then, Mary would have died, let us say, during time unit 31 and re-entered a new body — that of Lurancy Vennum — in time unit 33 — all very simple, straightforward, direct, and linear. But because of the time retraction abnormality, unit 31 coincided with unit 34 when Lurancy's physical body was already very much alive and established as a toddler — and presumably already occupied by another immortal entity since either conception or birth. Does that mean that both eternal spirit personalities were vying for control of that one physical body which was known by the name of Lurancy Vennum?

Another very convincing case involves Joanne McIver's recall under hypnosis of her previous life as Susan Ganier. At the time of her regression experiences, Joanne was living in Orillia about 130 kilometres north of Toronto. As Susan Ganier, she had been born in St. Vincent, Ontario, 150 kilometres from Sydenham, which later became Owen Sound. In July 1849, Susan had married Tom Marrow, a farmer. The ceremony had been conducted for them by a travelling preacher, Pastor

McEachern. The newlyweds had settled in Massie village, where they had been very happy together until Tom died as the result of an accident in 1863. Susan lived another forty years, dying in 1903. There were several interesting confirmations of Joanne's previous life as Susan to be found in various Canadian archives. The Ganier farm was recorded on an early map produced by the Ontario Department of Lands and Forests. Susan reported that she had a very good friend, Mrs. Speedie, who was in charge of the mail. She died in 1909 — just six years after Susan — and her tombstone is still readable in the cemetery at nearby Annan. The Toronto Department of Public Records was also able to verify the existence of several people whom Joanne had recalled when she was talking about her life as Susan Ganier. These included her friend Mrs. Speedie; a blacksmith named Robert MacGregor; William Brown, the miller; and Joshua Milligan, who kept the local store. The most telling evidence, however, came from Arthur Eagles, who was well into his eighties in 1969 when the details of Joanne's regression were published. Arthur remembered that when he had been a youngster, he had known the elderly Mrs. Ganier quite well.

Arnold Bloxham, who was well-known as a sensible and reliable hypno-regressionist, worked on one occasion with a helpful and responsive subject who clearly recalled her previous incarnation as a Jewish girl who had been killed in York during an outbreak of anti-Semitism there in 1190. During this tragic incarnation, under the name of Rebecca, she and her child had been trapped in the crypt of a little church and were unable to escape from a homicidal mob. On the tape that Bloxham recorded, the Jewish girl's re-enactment of her terror was charged with very realistic emotion. Professor Barrie Dobson of York University was consulted, as he was an acknowledged expert on the history of York in the twelfth century. Having listened carefully to the Rebecca tape, he gave his opinion that St. Mary's in Castlegate was closest to the description that she had given of the church in which she had been killed. The problem was that it had no crypt — and everything in Rebecca's account of her death depended upon a crypt being there.

Several months later, an amazing discovery was made. Artisans renovating St. Mary's Castlegate accidentally broke through into a crypt below the chancel. They reported seeing stone arches and vaults of a

Norman pattern — a style that existed well before the tragedy of 1190. Further discoveries of Norman and Anglo-Saxon stonework on the site leave no doubt that the archaeological evidence discovered after the hypno-regression supports the tragic story of Rebecca and her child.

A very different time anomaly was experienced by Major A.D. McDonagh while he was serving with the Indian Army along the Northwest Frontier. While riding across a range of hills close to the River Indus, McDonagh suddenly found himself surrounded by an ancient Greek army, which he later thought he might have been part of during an earlier incarnation. There was a great sense of sadness and grief among the men, and McDonagh soon discovered that they were mourning the loss of a brave and popular general — one of Alexander's senior officers. He went with a group of soldiers to a spot at the head of the valley and saw them looking at a newly inscribed grave marker. In his ordinary life, McDonagh could not read Alexandrian Greek, but he could read and understand this inscription that honoured and mourned a Greek general. History records that Alexander turned his attentions to India in 326 B.C and led a number of hard and bitter campaigns against fearless Indian warrior nations in the high hills around the Indus. One of the most noteworthy and memorable of these was the battle for the Indian stronghold of Massaga. Alexander himself led the campaign against the dauntless Ashvakas, and when their intrepid warrior chieftain, Assakenos, fell in battle, his equally courageous mother, Cleophis, rallied his troops and took command. Following her example, the Ashvaka women also joined the fight alongside their men. Vast amounts of emotional energy must have been generated from both sides during that grim, bloody, and protracted battle. Many experienced researchers into unexplained phenomena theorize that strong emotional energy can leave indelible psychic recordings at certain sites. Was it such a recording that helped to produce the apparent time slip that affected McDonagh?

The major was so intrigued by his strange experience with the Alexandrine troops that he felt compelled to return and explore the heavily wooded slopes where he was certain that he had seen the Greek general's memorial stone when it was newly carved. Returning with a party of Indian helpers, McDonagh cut his way through thick undergrowth to the spot where he was sure the ancient inscription was situated — and eventually he and his men located the stone. By now, it was badly eroded and faded, but recognisable fragments of the old Greek lettering remained.

How can McDonagh's evidence best be understood? Had he been drawn back in time because he was once an Alexandrian soldier who had fought in that part of India? Had time slipped there because of the vast emotional forces that had been released in the vicinity more than two millennia earlier?

His experience adds a new dimension to our considerations of time's mysteries and the strange tricks that time seems able to play on us from time to time. What if time has some curious power of attraction — something akin to electromagnetism in physics? The power of an electromagnet has to be switched on before its attraction can operate. Was it the emotional outpouring at the Battle of Massaga that put the switch into position? Or was it that an eternal spirit that had once occupied the body of a Greek soldier, and now occupied the body of Major McDonagh, was drawn back to that earlier, Alexandrian life?

Another fascinating case of being drawn back to a previous life was that of Joan Grant, the writer of *Winged Pharaoh.* Born in 1907, Joan was the daughter of wealthy parents and enjoyed a happy and privileged childhood, but even as a very small girl she would tell stories of the lives she had lived, describing them as things that had happened to her "before I was Joan."

During the First World War, Joan, still a young child, came down to breakfast having had a horrifying nightmare in which she saw soldiers being killed in action. In her inexplicable nightmare Joan had been with a soldier named McAndrew and had actually witnessed his death in action. She was able to describe his regimental badge with uncanny accuracy and sensed that it wasn't British. A soldier who was visiting her parents at the time listened to her attentively and sympathetically. He promised to make some inquiries, as he had realized from young Joan's description that the badge was Canadian; he later found out that a Canadian regiment had been involved in a night attack at the time Joan had been having her nightmare. During that fateful attack, a brave young Canadian soldier named McAndrew had died heroically in action.

Among other distinguished visitors to Joan's family home was C.G. Lamb, the brilliant professor of Engineering from Cambridge University. He and Joan got on well, because Professor Lamb was interested in the scientific aspects of psychical research. Joan's grandmother, Jennie Marshall, had been a superb musician — up to professional concert standards — but had died several years before Joan's conversation with Lamb. Knowing that he would not scoff at her, Joan told him that Jennie came and gave her music lessons. Then she played a piece that Jennie had taught her. Lamb was astonished. He had known Jennie Marshall in the past, and also knew that particular piece. He told Joan that only one manuscript copy of that music had existed, and that it had been given to Jennie years before by no less a person than the Tsar of Russia. Tragically, when Jennie had been diagnosed as terminally ill, she had burnt the manuscript along with several other unique pieces of music. Lamb told Joan that there was no way that she could have heard it in the normal course of events because it had been destroyed before she was born.

As a young woman of sixteen, Joan met H.G. Wells, who was just as interested in the scientific side of anomalous phenomena as Professor Lamb was. Wells also listened to her sympathetically and advised her to become a writer: good advice that she followed with great success.

Joan's psychic perceptiveness and sensitivity included psychometry. In 1936, she was given an Egyptian scarab that enabled her to envisage

Egyptian scarab.

scenes that she believed were part of a previous life that she had led in ancient Egypt as a woman named Sekeeta, who was the daughter of a pharaoh. She wrote these fascinating and compelling episodes from what she believed was her ancient Egyptian past life into her book *Winged Pharaoh*, which deservedly became a bestseller. Joan died in 1989.

Again, assuming that Joan's past-life experiences were real ones, how do they affect our concepts of time? If we pursue the reincarnationist argument that the immortal essence of her eternal inner self had once lived as Sekeeta in ancient Egypt — and that over the intervening centuries she had also lived through many other incarnations — we again seem to have a picture of unidirectional, arrow-line time as a medium for a series of consecutive incarnations. But how does this square with Joan's horrific nightmare of seeing the death of the courageous Canadian soldier McAndrew at the time when it was actually occurring? The experiential nature of time is one of its central mysteries: an innocent young girl whose body is safely asleep in a comfortable and secure English home seems to be sharing the horrors of a First World War night attack, hundreds of miles away in war-torn France. This resurrects the argument discussed earlier concerning the nonmaterial mind's apparent ability to experience another time in another place — or even the same time in another place.

No survey of the mysteries and secrets of time — and the curious tricks that time seems to play periodically — would be complete without

Egyptian chariot.

reference to American Edgar Cayce. Transparently sincere and trustworthy, Cayce was born on March 18, 1877, in Hopkinsville, Kentucky, and died on January 3, 1945, in Virginia Beach, Virginia. During his all-too-short sixty-eight years on Earth, Cayce made massive contributions to the health and welfare of thousands of his patients and clients. He was a pioneer of hypnosis and auto-suggestion as healing aids, and his researches into the mysteries of Atlantis and ancient Egypt were penetrating and significant. A firm believer in reincarnation, Cayce put forward several theories about previous lives in Atlantis.

Vast archives of his writings are preserved in the Association for Research and Enlightenment (ARE), which has its headquarters in Virginia Beach; there are similar Cayce research centres in over twenty countries worldwide. It is Cayce's firm belief in reincarnation and his theories about Atlantis that are most relevant here. He was a man of the highest moral principles, and the records of his many healings via hypnosis and auto-suggestion testify to his impressive powers, which were exclusively employed according to his high ethical standards.

Cayce's contribution to the investigation of the nature and meaning of time is inseparable from his case studies of previous lives in both Atlantis and ancient Egypt. According to Cayce, many reincarnated souls living in what was his contemporary world of the 1920s, 1930s, and 1940s were once Atlanteans. This again suggests the unidirectional, arrow-line model of time along which souls pass sequentially from the past to the present, and ultimately to one of a number of possible futures. Cayce was also a firm believer in free will. He was certain that genuine choice exists, and that the future is indeterminate until choices have been made in the present.

The gifted and mysterious Edgar Cayce.

His view of Atlantis — based on what he believed to be reincarnation readings — was that it had once been a gigantic continental land mass, whose exceptionally intelligent citizens enjoyed a high level of technology.

In two of our recent Dundurn books, *Mysteries and Secrets of the Templars* and *Mysteries and Secrets of the Masons*, we suggested that ancient, hidden Guardians at the back of both benign organizations were protecting humanity from sinister, secretive, and exploitative entities — and that these Templar-Masonic Guardians have been looking after us for millennia. Cayce's Atlantean information points in much the same direction. He writes of good forces in Atlantis, whom he refers to as Sons of the Law of One, and evil forces, whom he names the Sons of Belial. Coming as Cayce did from a devout Christian background, it is probable that his ideas about Belial were close to Paul's ideas in II Corinthians 6:15: "What harmony is there between Christ and Belial?" In other early Christian writings and some apocryphal sources, Belial is seen either as the father of Lucifer, or as a synonym for Satan. He is also described as a leader of the fallen angels who lost the war in Heaven and were duly expelled. In *The War of the Sons of Light Against the Sons of Darkness* that was found among the Dead Sea Scrolls, Belial is in charge of the Sons of Darkness. The book describes him as "Belial, an angel of hostility ... his purpose is to bring about wickedness ... all the angels that are with him are angels of destruction ..."

One of Cayce's most specific pieces of Atlantean information was that a mysterious blue stone would be discovered on an island in the Caribbean, and that it would have healing powers. In 1974 — almost thirty years after Cayce's death — a mysterious blue stone known as larimar (a form of volcanic pectolite) was discovered in Barahona in the southwest of the Dominican Republic in the Caribbean. Exponents of crystal healing credit larimar with healing powers.

In addition to his Atlantean revelations, Cayce also claimed to have memories of his life in ancient Egypt, where Atlantean learning and culture had been taken after the destruction of Atlantis itself. In his Egyptian incarnation, Cayce had been a benign priest and teacher named Ra Ta — almost certainly one of the ancient Guardians — and his later work in Virginia was a continuation and development of what he had accomplished in ancient Egypt many centuries before.

In any study of Cayce and reincarnation, it should be noted that water is a very prominent feature of Virginia Beach, with the mighty Atlantic to the east and Chesapeake Bay to the north. Edgar Cayce's proximity to the sea makes it look as if time anomalies and water are correlated yet again.

If Edgar Cayce cannot be left out of any comprehensive research work on reincarnation, neither can Bridey Murphy — or, rather, twenty-nine-year-old Mrs. Virginia Tighe, who recalled her past life as Bridey, apparently born in Cork, Ireland, in 1798. For just over a year, from October 1952 until November 1953, a hypnotist named Morey Bernstein worked with Mrs. Tighe, who had come originally from Madison, Wisconsin, and then moved with her family to Chicago. Mrs. Tighe had never been to Ireland during her twentieth-century life. Although investigated in depth, and questioned by a number of skeptics, the Bridey Murphy case stood up remarkably well. As in almost all reincarnation accounts, some events fitted so well with the "remembered" past that it seems impossible for them to have been put together by anyone who had not lived in that place at that time. The most intriguing detail in the Bridey Murphy case was her recollection of the Irish Uilleann pipes being played at her funeral in Cork in 1864, while her immortal essence waited in the limbo-between-lives that is part of reincarnation theory. These Uilleann pipes are the national bagpipes of Ireland, and were played at funerals in the nineteenth century because of their soft, sweet, plaintive notes. The air comes from a small set of bellows below the piper's right arm. The piper, therefore, does not have to blow to keep the bag inflated: some pipers can chat with other musicians as they play, or sing while providing their own accompaniment.

Another very significant case of reincarnation — in fact, several very significant cases — were brought to light, analyzed, and examined by Dr. Arthur Guirdham, a trustworthy and reliable academic psychiatrist. One of his patients had experienced terrifying nightmares for several years before coming to Guirdham for help. Her revelations about the persecutions of the Cathars of the Languedoc in southwestern France during the thirteenth century were so historically accurate and compelling that it seems highly probable that she had actually experienced a previous incarnation as a Cathar. Co-author Lionel was certainly warmly and favourably impressed when he met and interviewed Dr. Guirdham. A highly intelligent doctor of medicine and a leading psychiatrist, he would be the ideal man to separate elaborate fantasy from genuine paranormal experiences. In Dr. Guirdham's view, his patient was recalling detailed memories of a tormented past life during which she was brutally persecuted for her Cathar faith. His investigations led him to conclude that not only had his patient experienced a previous life in the French Midi during the thirteenth century, but several other twentieth-century friends and colleagues had as well, and it seemed feasible to suggest that some souls, perhaps, travelled in groups.

Evidence for reincarnation of the type provided by Guirdham (in his books about reincarnated Cathars), by Edgar Cayce, and by Virginia Tighe is worth careful consideration, but in addition to the case for reincarnation, the concept of déjà vu also needs to be investigated thoroughly in connection with the mysteries and secrets of time.

The term *déjà vu* comes from the French "already seen"; it is also referred to in psychology as paramnesia. Another form of paramnesia, referred to clinically as reduplicative paramnesia, is the feeling that a location exists in two places at once. Arnold Pick, a Czechoslovakian neurologist, had a patient in 1903 who was convinced that she had been moved from his clinic in the city centre to an identical one in a city suburb with which the patient was familiar. She rationalized her belief in this non-existent second neurological clinic by convincing herself that Pick and his staff colleagues worked in both locations. In wartime, a number of clinicians reported cases of wounded soldiers — especially those who had head injuries — who believed that the distant military hospitals in which they were being treated were actually located in their own hometowns. Researchers into the anomalous and paranormal have suggested an alternative explanation for reduplicative paramnesia: perhaps there are mental states in which the mind is aware of two or more parallel time-tracks. This would, of course, support a theory of bifurcated, coexistent time paths rather than the traditional, unidirectional arrow of time.

The widely accepted understanding of déjà vu is that it is the feeling that the observer in a new situation has actually been there before or seen it before. The phrase was coined by Émile Boirac (1851–1917), a French researcher into the anomalous. His book was entitled *L'Avenir des Sciences Psychiques* (*"The Future of Psychic Sciences"*). As Boirac explained, this sense of déjà vu is characteristically accompanied by a feeling that something strange, weird, and eerie is happening to the witness.

Déjà vu has been subdivided into three categories: déjà vécu, déjà senti, and déjà visité. The first one is déjà vécu ("already lived through"). Charles Dickens describes this feeling very perecptively in *David Copperfield*: "We have all some experience of a feeling, that comes over us occasionally, of what we are saying and doing having been said and done before, in a remote time — of our having been surrounded, dim ages ago, by the same faces, objects, and circumstances — of our knowing perfectly what will be said next, as if we suddenly remember it!" Another useful way of categorizing déjà vécu is to regard it as involving more than sight alone: the whole experience seems familiar to the observer — everything is just as it was before.

The second type, déjà senti ("already felt"), is a mental rather than a physical sensation. The subject who is going through a déjà senti phenomenon tends to describe it as a feeling or a state of mind, perhaps just a train of thought. The subject is then aware that those thoughts or feelings were interrupted for some reason, and then came back accompanied by recognition or a sense of familiarity.

The third type of déjà vu is déjà visité ("already visited"), and the subject experiencing it exhibits an extremely detailed and accurate knowledge of the place in question. Frequently, persons describing déjà visité can find their way around an old hall, palace, or castle as though they had lived there in the past.

Paranormal explanations for all three types of déjà vu include out-of-body experiences and reincarnation, and déjà vu is not easy to disregard when considering the nature of time.

Academic psychological and neurophysiological research has led to suggestions that the déjà vu phenomena are anomalies of memory. The mind thinks that something is being recalled when it is not — memory is wrongly claiming that the new inputs were there already. Some psychologists might suggest that the difficulties described as déjà vu occur when there are problems at the interface of short-term and long-term memory. Short-term memory is concerned with what is immediately present or only just past; long-term memory is concerned with the more distant past. If there is a problem at their interface, the subject may feel as if the immediate (short-term memory) data belongs in the past — and, therefore, he has encountered that data before.

Clinical theories have been advanced suggesting that some déjà vu experiences may be associated with schizophrenia and temporal lobe epilepsy. However, it has to be remembered that reputable research findings suggest that over 70 percent of perfectly healthy people interviewed reported having experienced déjà vu.

Pharmacological research highlighted a case of déjà vu where the subject had taken a mixture of two drugs to relieve his influenza. The pharmacological researchers examining this data concluded that the déjà vu experiences in this particular subject had been triggered by the effects of the anti-influenza drugs on the mesial temporal areas of the subject's brain.

Another scientific theory that attempts to explain déjà vu phenomena focuses on the possible effects of similar memories. If, for example, the subject has actually visited an old motte and bailey castle of the early Norman type, and now sees another one that closely resembles his true memory of the castle that featured in his earlier visit, then, according to this explanation, the subject may well report a déjà vu experience.

Some interesting neural theories have also been advanced to explain déjà vu. One of these suggested that if neuronal firing was mistimed in any way, the brain might be tricked into thinking that the information it was receiving along the faulty neuronal pathway had already reached it at some time in the past: hence the déjà vu experience. Similar neural theories concentrated on the eyes and suggested that if the signal along one optic nerve was faulty, so that its message reached the brain a little *after* the first signal from the efficient eye had been received, then the brain would again be tricked into thinking that the new data coming along the aberrant optic nerve was in fact data that had already been received — hence the feeling of déjà vu. Mistimed firing between the left and right lobes of the brain might have similar déjà vu effects according to some other neural theorists.

For parapsychologists, there is a strong nexus between reports of déjà vu and cases of precognition, especially clairvoyance, clairaudience, and what appears to be precognition. Other theorists wonder whether the memory of a particularly vivid and stirring dream may cause the dreamer to think that the dream scenario was real. If that location is then visited for the first time, there will be a sense of déjà vu because the vivid dream was indistinguishable from reality.

Other researchers have referred to the opposite phenomena, which is categorized as jamais vu ("never seen"). Jamais vu occurs when a familiar scene or a well-known friend or family member is not recognized. Clinical research has demonstrated that this is most likely to occur when the subject is overworked, stressed by the pressure of too many simultaneous demands, or exhausted for some other reason. Like déjà vu, jamais vu produces a weird, eerie sensation in the subject who is experiencing it.

Presque vu ("almost seen") refers to the subject having an intense, eerie feeling that he is on the verge of some great new discovery. The answer to some very important question is hovering just beyond the grasp of consciousness. The subject feels tense and excited — and at the same time very frustrated — because the very important thought that they are sure is there refuses to emerge.

The most ironic of the déjà vu experiences, and their closely associated mind states, is known to French psychologists as *l'esprit des escaliers* ("staircase wit" or "staircase humour"). This refers to what happens to almost everyone periodically: we think of the ideal reply, but too late. The brilliant riposte comes to us a full two minutes after we needed it. It ought to have flown to us on the wings of a diving hawk or in a single lightning flash of inspiration. It arrived instead very much too late and out of breath from climbing the staircase!

Whatever déjà vu may be, it cannot be ignored during research into the mysteries and secrets of time. If a déjà vu witness is certain that he has never visited a place before, but then finds its details are strangely familiar, then one possible explanation of the phenomenon is an anomaly in time.

During one of co-author Lionel's broadcasts on George Noory's very popular radio show *Coast to Coast* in the United States, the subject of déjà vu came up and resulted in an extremely interesting and relevant e-mail from Michael S. Ramirez, which reached the authors after the show. Ramirez was serving on board the *Raleigh* in December of 1966, and was in St. Croix with two shipmates, R.A. Hanson and A.C. Woodward. The men were feeling hungry, but none of them knew the area and had never visited it before. Ramirez possesses several remarkable psychic talents, and using one of them — which researchers might describe as something close to the déjà vu mind-state — he conjured up a clear mental image of a side street near the Cavanaugh where they would find a small shop that was adjacent to a good restaurant. There was also an excellent swimming pool and an attractive beach nearby. Following these directions, Ramirez, Hanson, and Woodward found everything there just as Ramirez had seen it. Even more intriguing was that a very attractive girl, Trudy Meyer, met them in the shop, smiled in recognition, and told them that her father would be there to greet them in a minute. It was as if she and her father already knew Ramirez, although he had never been to their shop before. The authors are very grateful to Michael Ramirez for sharing this experience, and for kindly allowing us to include it here.

Two strange time adventures that are very difficult to explain concern Sir Robert Victor Goddard (1897–1987), formerly Wing Commander Goddard of the Royal Air Force, as he was in 1935 when he became involved in the first episode. Goddard is an ideal — and very prestigious — witness. He joined the Royal Navy in 1910 and served with distinction in the RN Air Services, later transferring to the RAF. He was Deputy Director of Intelligence at the Air Ministry in 1938 and 1939, and served with the British Expeditionary Force in France in 1939. He was also in charge of the highly effective New Zealand Air Force in the South Pacific in 1941. From 1946–1948, he served as the RAF representative in Washington, eventually retiring in 1951.

The start of Goddard's 1935 time adventure occurred when he had been sent to inspect a disused First World War airfield at Drem, not far from Edinburgh, and had flown there in his Hawker Hart biplane. What he saw at Drem was a depressing landscape of dereliction and dilapidation: hangars in need of repair and cracked tarmac clearly looked as if the

The Hawker Hart biplane in which Wing Commander Goddard apparently flew through time at Drem.

airfield had not been used for years. Cattle were grazing there as if it was a familiar, peaceful, and long-used pasture as far as they were concerned. Later that day, as Goddard resumed his flight, the weather deteriorated badly. Skilful and experienced pilot though he was, the weather forced his Hawker Hart down sharply, and Goddard only just managed to control her a few feet from the ground. Flying back towards Drem to make sure that he really was where he thought he was, Goddard saw that the airfield had undergone an incredible change since he had seen it earlier that day. The grazing cattle had gone. The grass was neatly trimmed, and the whole place was in excellent repair. There were brightly coloured yellow planes on the ground, and mechanics in blue uniforms were attending to them. Even more surprisingly, the torrential rain and storm-force winds had abated: Drem airfield was now shimmering in bright sunlight — although something about the sunlight seemed odd, almost unreal, rather like stage lighting on a vast scale. Although Goddard was flying low over the airfield none of the groundcrew seemed to be aware that the Hawker Hart was just above them.

Goddard was more than a little surprised by the whole episode — not least because in 1935 RAF planes were normally silver or aluminium coloured, and engineers wore traditional khaki protective clothing. The Second World War broke out four years later, and Goddard had to make another visit to Drem; the restored and refurbished airfield was now identical to the one that he had seen in his "time slip" vision in 1935. The planes were painted yellow, and the engineers wore blue overalls.

In his second story, Goddard (by then Sir Victor Goddard) was at a party in Shanghai after the Second World War when he overheard someone nearby saying that he had just heard of Goddard's death! Sir Victor turned round in amazement and recognized the speaker — who also recognized him. The man who had been so sure that Goddard was dead was British Royal Naval Captain Gerald Gladstone. He apologized profusely and explained that the misinformation was the result of an all-too-vivid dream, which had led him to believe that Sir Victor had been killed when a Dakota had crashed. The details of the dream included three civilian passengers — two men and a woman — who had all emerged unhurt from the wreckage: Sir Victor, unfortunately, had not. Goddard was due to fly in a Dakota shortly, but there were no civilians scheduled to join him. However, shortly before takeoff, the Consul General told Sir Victor that he had been ordered to go to Tokyo as soon as possible, and begged a lift for himself and his female secretary. Goddard felt that he could hardly refuse. Then a male journalist from the London-based *Daily Telegraph* also begged a lift to Japan. It was not lost on Sir Victor that he now had three civilians with him — one female and two males — exactly as Captain Gerald Gladstone had seen in his dream. The weather deteriorated just as it had in the dream, and the plane had to make a crash landing on a rocky island. There the dream and reality parted company: everyone — including Sir Victor — survived.

General Toutschkoff was less fortunate. Several weeks before Napoleon invaded Russia, Toutschkoff's wife had a recurring dream. She was in an unknown inn in an unknown town when her father came into the room, holding her son's hand. Her father told her with great sadness that her husband had been killed in action at a place called Borodino. When the tragedy actually happened a few months later, General Toutschkoff was commanding Napoleon's Army of the Reserve. His father-in-law came into Mrs. Toutschkoff's room with his grandson and gave his widow the tragic news of her husband's death in battle: it was the room she had dreamt of, in the inn she had dreamt of, in the town she had dreamt of.

What — if any — is the explanation of the differences between apparent glimpses of the future that are fulfilled exactly, that are fulfilled in part, and that are not fulfilled at all? Suppose that there are *ripples* in time — some very energetic, others so gentle and smooth as to be almost imperceptible? Just as sunspots vary in their size, duration, and intensity, so these time ripples may do the same. Consider the likelihood, as has been suggested earlier, that we are all free beings and that we do have genuine choice. Each moment of our lives lays a fresh choice before us. Depending upon what we decide, so our future will vary. Our present decisions structure the track along which our consciousness will travel in the future. Perhaps what we do *not* decide becomes a subordinate probability track, an inferior World of If; but just because we chose not to enter it, we do not deprive it of existence. When time is running smoothly and with very few, gentle ripples, it may be comparatively easy to "see" these probability tracks, like reflections in clear, smooth water. Or perhaps we experience them in some other way rather than sight? When time is disturbed by powerful forces that create turbulent ripples in it, the "reflections" that reach our experiential world from those other probability tracks will vary significantly from what actually happens here because of the distortion created by the ripples. (It was Charles Fort who once said with great wisdom and insight: "Tread carefully on the thin crust that we call *reality.*") Sometimes a dream or a psychic prediction will turn out to be close to actual later experience. More often than not, it won't.

The case of the fictional *Titan* and the tragically real *Titanic* disaster illustrates the point. In 1898, a writer named Morgan Robertson (1861–1915) wrote a story called *Futility*, which after the *Titanic* disaster of 1912 was re-titled *Futility, or the Wreck of the Titan*. In Robertson's book, his fictional ship, the SS *Titan*, described as the fastest, the safest, and the most luxurious liner afloat, set out on her maiden voyage in 1898. The *Titan* was 70,000 tonnes. The real *Titanic* was 269 metres long, with a 28-metre beam. She was 18 metres from the waterline to her boatdeck, and her gross registered tonnage was 46,328 — a long way below that of the fictional *Titan*. Both ships were driven by three screw propellers. The fictional *Titan* went down in April 1898. The real *Titanic* struck the iceberg and sank on April 14, 1912. Both ships were heading across the Atlantic from the United Kingdom to the United States when each struck an iceberg and sank. The majority of the fictional *Titan*'s 2,500 passengers died in the disaster, due in no small measure to the very few lifeboats — only twenty-four. The real *Titanic* had only twenty lifeboats. In Robertson's 1898 story he refers in detail to the watertight doors of the *Titan*:

> From the bridge, engine-room, and a dozen places on her deck the ninety-two doors of nineteen water-tight compartments could be closed in half a minute by turning a lever. These doors would also close automatically in the presence of water. With nine compartments flooded the ship would still float, and as no known accident of the sea could possibly fill this many, the steamship *Titan* was considered practically unsinkable.

The actual *Titanic* had sixteen watertight compartments, held by a magnetic latch.

It was not only Morgan Robertson's 1898 story that seemed to have encapsulated a "psychic reflection" via a time ripple.

William Thomas Stead (1849–1912) was a crusading English journalist, famous for the notorious Eliza Armstrong case that was part of his campaign against child prostitution in Victorian England. Despite several warnings from psychics, and Stead's own strange, prophetic dreams of disaster at sea, he was one of the celebrities who drowned when the *Titanic* went down. One of the psychics had told him: "I see you struggling in the sea with 1,000 others … none escape, yourself included."

A business executive named Middleton had booked on the *Titanic* and, like Stead, had had a number of disconcerting dreams about a disaster at sea. In Middleton's dreams, he saw the *Titanic* floating keel upwards, with the doomed passengers and crew struggling in the icy water around her. Fortunately for Middleton, his New York business conference was called off, and he was able to cancel his booking on the *Titanic*. Another heeded warning saved the life of a naval engineer named Colin Macdonald. He was quite sensitive to strange, psychic feelings, and felt very uneasy about sailing on the *Titanic*. The job of Second Engineer, a very prestigious post on such an enormous ship, was offered to Colin on three occasions — and each time he turned it down because of his psychic forebodings. The engineer who accepted the post died when the *Titanic* was lost. Did Macdonald's strange feelings of impending disaster come to him because he had seen a time ripple reflection of the loss of the *Titanic*?

Time seems to play a wide variety of tricks that hardly seem compatible with one another. It appears to move its own goal-posts so that mere human players cannot win the mind-games it invites us to join. But if we continue to focus on it sharply enough, and to study it for long enough, some of those strange tricks will eventually begin to make sense — and we shall start to see through the philosophical and metaphysical camouflage with which time surrounds itself. Sooner or later, the rigorous application of objective science will bring us closer to the true mysteries inside time's innermost secret cabinet.

Chapter 10

DID THEY SEE THE FUTURE?
PROPHETS AND SEERS

AN EXAMINATION of the writings and utterances of people like Nostradamus, Mother Shipton, the Brahan Seer, and hosts of other prophets and foretellers of future events immediately raises the question of what it is that they are actually seeing and hearing. If it is suggested that they are seeing a future that is as rigidly fixed as the past, then the horrendous implication is that we have no choice over our lives. There is no room for ambition, for ethics, or for morality. None of us can be held responsible for what we do, or for the choices we make. Good and evil as philosophical abstractions may still exist, but good and evil *people* do not. If we are simply automata under the control of fate, destiny, Lady Luck, or any other controller, then we have never done anything for which we can be thanked and praised, nor for which we can be hated and blamed.

All our experience and common sense tell us that life is not like that and *cannot* be like that. We *do* have choices to make moment by moment. I can give a dollar to the homeless man shivering in a shop doorway, or I can pass by as if I hadn't seen him. I can shrug and laugh it off when some idiot who thinks he's a smarter driver than the rest of us cuts up arrogantly and illegally on the wrong side to gain a place or two in the traffic. Alternatively, I can succumb to what is popularly known as "road rage," leap out when he stops at the next red light, put my martial artist's fist straight through his windscreen, and punch his head. The consequences of my choice at that instant will have a profound effect on my future and his: I go to prison for assault; he goes to hospital — or the mortuary, depending on how far back his head went when I hit him. So I choose to laugh and shrug it off: he goes wherever he was planning to go, and I go on to the university to deliver my lecture.

Then we enter the mysterious realm of probability tracks and the Worlds of If. What if I'd chosen the *wrong* alternative? Does his premature death, and my lengthy imprisonment for causing it, have some sort of quasi-existence elsewhere?

When an artistic genius is planning a picture, she considers dozens of alternatives subjects, compositions, shapes, forms, colours, lights, and shades. One picture is finally painted: a hundred others are merely thoughts in the artist's mind. The final picture that hangs on the wall of the National Gallery is the experiential reality shared by the Gallery staff, the artist, and the viewers. Vestiges of the pictures that were not created may linger on in the artist's mind. These are merely probability tracks, yet they do, arguably, still have a kind of existence. Suppose that, after a time, the artist decides to use one of those plans to create another, similar, picture. What was formerly a mere probability track now becomes an experiential reality: it has made the transition from quasi-existence into what we call reality (without always understanding what we mean by "reality").

Take it a stage further. The artist has friends and colleagues in the art world. Perhaps she has made a few rough sketches to discuss with them before committing herself to the final picture. Those friends, colleagues, and advisers have had access to the probability tracks. In this analogy, can we liken those people to the seers, prophets, and foretellers of future events? The real, final, completed picture is in no way fixed and inevitable just because the rough sketches have been seen and the artist's ideas have been discussed. It will not become a fixed experiential reality until the artist has put paint to canvas, brushstroke by brushstroke.

Many researchers into the paranormal and anomalous have been intrigued when friends and colleagues have shown them pictures of inexplicable glowing spheres that seem to have the ability to float in and out of normal reality. The authors are very grateful to one such friend for giving permission to reproduce here some intriguing pictures that she took of these enigmatic, transient spheres.

Numerous explanations have been suggested. Some researchers have hypothesized that the spheres are of extraterrestrial origin and may be able to acquire information and transmit it back to their alien senders far away. Psychic theories include the idea that the spheres are disembodied spirits, consisting of pure, non-physical mind or personality. Other researchers have looked into the idea of astral projection and out-of-body experiences. The spheres may be visitors from other times, from other dimensions, or from other probability tracks — travellers from the Worlds of If — having the quasi-existence of what *might* have been in a future that was not fulfilled: the very unwelcome future, for example, that would have become reality if co-author Lionel had chosen the road rage option! Did prophets and seers, like Mother Shipton, come into contact with similar globes?

Ursula Sontheil, the Yorkshire Wise Woman and Prophetess of Knaresborough, was known as Mother Shipton. She was born in the closing years of the fifteenth century to a young orphan girl called Agatha Sontheil. Strange stories accompanied the birth itself, which

A genuinely mysterious orb picture, provided for us by the witness who took the photograph. Do these mysterious glowing spheres come to Earth via time portals?

During research into how such pictures can be the result of misunderstandings and misinterpretations, this "orbs-around-the-chandelier" photograph was created by the authors using a fine spray of water droplets.

traditionally took place in a cave beside the River Nidd, which still attracts tourists who are fascinated by Mother Shipton's prophecies. The women attending Agatha at the time of Ursula's birth said that as the child emerged into the world there was a terrifying crack of thunder and an ominous smell of sulphur. There were even rumours that Agatha had been seduced by an incubus (a male sexual demon) rather than by a man.

Various theories about her father's identity have intrigued historians for centuries. The first suggestion is that he was well up in the church hierarchy. It is known that the Abbot of Beverley Minster came in person to baptize baby Ursula — a very unusual honour. Was the Abbot himself the father, or was he trying to make amends for the irresponsible behaviour of one of his men? Not only did the Abbot carry out her baptism, he kept a benign and caring eye on the child for several years afterwards.

Another theory laid the responsibility on an itinerant troubadour passing through Knaresborough. Perhaps Ursula's father was a soldier of fortune, a mercenary warrior between battles, or even a knight or nobleman. There is some uncertainty about what happened to Agatha within a year or two of the baby's birth. There is evidence that she died young and that a local nurse then took care of the tiny orphan. It has also been suspected that Agatha was deliberately done away with in order to protect the mysterious father's identity. In another version of Ursula's early infancy, it was Agatha who gave the baby to the nurse, and then went off to a convent for the rest of her days. A strange tale surrounds Ursula while she was in the nurse's care: both she and the nurse were found in the chimney of the nurse's cottage! Ursula was still in her cradle, and neither she nor the nurse seemed to have any visible means of support: they just hung there in the chimney. Stories like that — no matter how wildly exaggerated they were — led to all sorts of sinister speculations about the unknown father's identity. Some village gossips said that he must have been a demon, perhaps even Lucifer or Satan himself.

There are other mysteries surrounding the child's name, Ursula. *Ursus*, from which Ursula is derived, means "the Bear." The

The petrifying stream at Mother Shipton's Cave in Yorkshire. Items hung in this water become covered in stone.

The River Nidd beside Mother Shipton's Cave.

famous fifth-century war leader, known romantically as King Arthur, was probably known as Art or Arth in Wales. The Welsh *Art* ("bear") and the Brittonic *ur* ("man") would have given a name meaning "the man-of-the-bear," or simply "the bear-man," a tribute to his size and strength and to his protective role against waves of Anglo-Saxon invaders. There are other experts who link King Arthur's name to the star Arcturus. Arcturus is taken from Classical Latin. By the fifth century, when Late Latin was in use, the name would have changed to Arturus. This star was associated with the constellation of the Great Bear.

Was the mysterious visitor to Knaresborough who fathered Agatha's child a descendant of one of Arthur's knights? As far as the magical side of young Ursula's life was concerned, could there have been any connection with Merlin, the Bear-King's loyal wise-man and magician?

Is it possible that Ursula's father called on the Abbot to help look after the child when he could not? Perhaps the King had summoned him to an overseas war. The king at the time of Ursula's birth was Henry VII, the Welsh Tudor monarch, who would have had massive respect for Arthur of the Bear Clan. Was that absent father also a member of the noble Bear Clan, and was the obliging Abbot yet another member, who felt that he had to assist a brother Bear Clan member by looking after his child?

King Arthur.

Wherever the finances came from, Ursula was educated at a good school near Knaresborough. The other children laughed at her and made her life a misery because of the arthritis she suffered from, but the teachers were kind, helpful, and understanding and enabled her to make the most of her lively mind and high intelligence. It was while she was still at school that her unusual ability to communicate sympathetically with birds and animals was first noticed.

At the age of twenty-four, Ursula met, fell in love with, and married a local carpenter, a good-hearted and caring man named Toby Shipton, who was not in the least put off by the arthritis that had made Ursula's classmates laugh at her when she was a child. It was after their marriage that Ursula became known as Mother Shipton.

An interested chronicler named Joanne Waller, a contemporary of Ursula's, wrote down many of Mother Shipton's strange prophecies. Both ladies later died in the same year, 1561, at the start of the reign of Queen Elizabeth I.

One of the darkest and grimmest of Mother Shipton's strange, rhyming prophecies concerned a young nobleman who was very deeply in debt and was being pressed hard by his creditors. He came to ask Mother Shipton how long his father was likely to live, because he thought that if he could tell his creditors that he would soon be inheriting enough to pay them, they might be willing to give him more time. Disgusted by the young man's request, Mother Shipton refused to answer his questions. Shortly afterwards — possibly because of the stress to which his creditors were subjecting him — the boy became ill. His father, deeply concerned about his son's prognosis, came to consult Mother Shipton, who said: "Those who gape out for others' deaths / Their own, unlooked for, comes about. / Earth he did seek; ere long, he shall have / Of earth his fill; within his own grave."

Shortly afterwards, the young man's illness took a turn for the worse, and he — who had "gaped out" for his father's death — died himself.

Her wider and more significant prophecies often concerned national and international issues. The marriage of Henry VIII and Anne Boleyne, sometimes spelled Bulloigne, or Bullen, was thought to have been forecast in Mother Shipton's couplet: "When the Cow doth ride the Bull, / Then, O Priest, beware thy skull."

Henry's armorial bearings included a cow, and the Boleyne insignia included a black bull's head. Shortly after their marriage in 1533, Parliament passed the Act of Supremacy, which placed Henry at the Head of the English Church with power to confiscate monastic property. Anne was executed in 1536, and Henry's actual confiscation of monastic property lasted from 1538 until 1541. The monastic clergy certainly had every reason to beware of Henry and his equally greedy and unscrupulous henchmen during those brutal and disastrous years. Henry's daughter by Anne grew up to become Queen Elizabeth I, and of her Mother Shipton wrote: "A maiden Queen for many a year / Shall England's war-like sceptre bear."

Mother Shipton also seems to have foreseen Drake's contribution to the defeat of the Spanish Armada: "The Western Monarch's wooden horses / Shall be destroyed by the Drake's forces."

The Great Plague and the Great Fire of London also seem to have appeared in her visions: "Triumphant Death rides London through / And men upon the housetops go ..." During the rapidly spreading fire, anxious Londoners climbed up on to their rooftops to try to see which way the flames were spreading. Samuel Pepys wrote in his diary for October 20, 1666, that Sir Jeremy Smith had said, "now Shipton's prophecy is out" (using "out" in the sense of "fulfilled").

One of Mother Shipton's personal prophecies concerned Cardinal Wolsey. There was deep enmity between them: she referred to him as "the butcher's boy" because his father had owned a butcher's business in Ipswich in Suffolk, and as "the mitred peacock." Wolsey swore that he would have her burnt at the stake as soon as he reached York. Mother Shipton replied that he would *see* York but never *enter* the city. Wolsey's party reached Cawood Castle, and the cardinal climbed the tower to get a good view of York. That same evening, as Wolsey was feasting in Cawood, the Earl of Northumberland arrived and arrested him on Henry VIII's orders. Wolsey was charged with treason, and in Henry's day the accused's chance of emerging alive was very slight. Wolsey, more fortunate than many of Henry's enemies, died of what appeared to be natural causes on the way to London to stand trial. It is possible, however, that he committed suicide by poisoning himself.

Another of Mother Shipton's warnings was issued to the Lord Mayor of York, who took up residence in Minster Yard. She said: "When there is a Lord Mayor living in Minster Yard in York, / Let him beware of a stab..."

Thieves stabbed and robbed the Lord Mayor one night, and he died of his wounds.

Mother Shipton's prophecies were well known by the Patagonian Welsh, which might have something to do with her possible connection with Arthur's Welsh Bear Clan. The Patagonian Shipton verse could have been a prophecy of the Falklands War:

In time to come in future
Our land shall have two women rulers.
One of our children will be contested
By a child of Spain ...
In this time, blood will be shed,
Yet shall our child stay ...

Researchers, including David Toulson, have seen ingenious connections between Shipton's words and the Falklands engagement. Britain's two female rulers at the time were Queen Elizabeth II and Prime Minister Margaret Thatcher; the British "child" could have been the Falkland Islands, and the Spanish "child" could have been Argentina. Blood was shed in the battle for the Falklands — and Britain retained the islands.

Her predictions of future technology were also impressive:

Carriages without horses shall go
And accidents fill the world with woe.
Around the world thoughts shall fly
In the twinkling of an eye ...

She seems here to be forecasting cars and trains, and there's nothing as effective as the Internet, e-mails, and satellite digital technology to send human thoughts around the world almost instantaneously — literally "in the twinkling of an eye," just as Mother Shipton said.

Michel de Nostredame was born on December 14, 1503, in St. Rémy, Provence, France. His mother, Renée, was warmly loving, and a dedicated and effective homemaker. His father, Jacques, was a good provider: he practised as a notary and earned a satisfactory income. Young Michel grew up in ideal circumstances: he was surrounded by love and comfort, and nourished by good Provençal cuisine. Meals were extensive family affairs, and the boy's mind was stimulated by interesting conversation. His academic grandfathers, Jean de St. Rémy and

Pierre de Nostredame, taught him Latin, Greek, and Hebrew, plus a wide range of other disciplines. Michel learned medicine, herbalism, literature, and history from them — as well as the secrets of the Jewish Kabbalah (or Cabala) and alchemy.

Survival amidst the savage religious bigotry of fifteenth- and sixteenth-century Europe depended upon conforming to the form of Christianity acceptable to the Inquisition, so Nostradamus's wise

Bust of Nostradamus.

and prudent family conformed. In secret, in the security of their well-guarded homes, they continued to practise their age-old Jewish faith, and to uphold its wisdom, faith, and culture.

Michel's grandfathers had been responsible for his parents' meeting in the first place. The two highly intelligent and expert physicians had the care of liberal and tolerant old King René of Provence — rightly remembered in history as René the Good. They also cared for his son,

Co-author Lionel in the Nostradamus Museum in Salon, Provence, studying historic documents with the curator.

the Duke of Lorraine and Calabria. In the circumstances, there was a high probability that Jean's daughter would marry Pierre's son. The happy, golden age of René ended with his death in 1480, and it was unfortunate that the throne passed to Louis XII of France, who lacked René's enlightenment. During the reign of Good King René art, theatre, and the superb products of the Muscat grape had been warmly appreciated by his grateful subjects.

Michel's studies of grammar, philosophy, and rhetoric began in Avignon when he was fourteen. It was his misfortune to find that the college authorities were traditional, conventional, and authoritarian priests, whose ignorance led them to look with disfavour on Michel's interests in the occult and in astrology. Nostradamus was also intelligent enough to realize that Copernicus was right about the relative roles of the sun and planets in the solar system, and argued the point fearlessly, even though it was not a safe time to become involved in scientific-religious controversies. This was the period when Martin Luther (1483–1546) was busily nailing his Declaration to the Wittenberg church door, and John Calvin (1509–1564) was about to launch his narrow, fatalistic Puritanism on an unsuspecting world. In due time, Nostradamus — who detested Calvin's narrow religious bigotry — would refer to his theology and religious teaching as "that loathsome odour."

Politically astute old Grandfather Pierre called Michel home and gave him some very shrewd advice about when it was appropriate to speak out and when it was *not*. Much of the religious misery of the fifteenth and sixteenth centuries arose because Christians at that time had an irrational dislike of Jews, whom they regarded as the murderers of Christ — conveniently forgetting that Christ himself was Jewish. Tomas of Torquemada (1420–1498) was one of the most active inquisitors of all time, and was responsible for thousands of deaths. He launched his venom against Spanish Jews in particular. Many had fled to the safety of liberal Provence and the protection of René the Good. Under the rule of Louis XII, however, Provence was far less safe than it had been, and wise old Pierre had no difficulty in persuading his intelligent and perceptive grandson that it was prudent to keep a low profile and walk on eggshells. As a teenager, Nostradamus had expressed keen enthusiasm for a career as a professional astrologer, but he listened again to his grandfathers' wise advice and studied medicine at Montpelier University. Starting his course in 1522, Nostradamus qualified in 1525, having already learned far more from his grandfathers than from the professors at Montpelier. He left Avignon with alacrity, and went off to practise his own kind of medicine in his own way. European cities of the time were overcrowded and unhygienic. Southern France was plagued by *le Charbon,* a hideous form of bubonic infection.

Many of Michel's patients were cured by his advocacy of clean drinking water, fresh air, and the administration of therapeutic herbal remedies that he had acquired from his grandparents.

Did Nostradamus know about pathogenic micro-organisms because he had seen various future probability tracks? There is some evidence that he actually named Louis Pasteur in his coded verses: is it remotely possible that during one of his strange visions Nostradamus had seen Pasteur in action? The tragic irony of Nostradamus's medical career was that — despite saving many of his patients — he was unable to save his own beloved wife and children, all of whom died of plague. Heartbroken, Nostradamus became an itinerant healer for the next few years, and his prophetic powers grew during this period.

On one occasion, he encountered a group of Franciscan friars on a narrow, muddy road, and stepped respectfully aside to let them pass. Suddenly he fell to his knees in front of a young brother named Felice Peretti, announcing that this young man would one day become Pope — and forty years later that young Franciscan was elected Pope Sixtus V.

After his sad and lonely years of wandering, Nostradamus came to Salon in Provence, where he met and married a beautiful young widow, Anne Posart Gemelle, who was also wealthy enough to see that they both lived in comfort. They lived happily in her beautiful home in the Rue de la Poissonerie, where Nostradamus used the top floor as his study. It was in this house — now the Nostradamus Museum in Salon — that many of his most famous and enduring prophecies were written. He began his trips into what seem to have been future possibility tracks — the Worlds of If — by staring at a thin flame and emptying his mind. He then sat on a brass tripod and gazed into a brass bowl filled with water, drifting into a trance-like state. Once he had achieved that, he began to hear messages and to see successions of mysterious pictures and visions in the water. The technique is usually referred to as hydromancy, and is similar to catoptromancy, which employs mirrors instead of water. J.R.R. Tolkien, in *Lord of the Rings*, refers to the Mirror of Galadriel, through which the Elven queen is able to show Sam Gamgee, one of the Hobbit heroes, scenes from his home in the Shire.

Nostradamus, a good and caring man, found himself confronted by an ethical and moral dilemma. On one hand, he felt that it was his duty to share his visions with others so that certain dangers might be avoided; on the other hand, he was afraid that to be known as a seer would invite persecution and serious charges of heresy and witchcraft. The solution he found was to use obscure codes and to hide his visions in convoluted verses. It was a case of "He who has eyes to see, let him see. He who has ears to hear, let him hear," as the Bible says. Nostradamus published his first Almanac containing twelve mysterious quatrains in 1550, and brought one out every year for the rest of his life.

Co-author Lionel beside a waxwork of Nostradamus in the Nostradamus Museum.

One of Nostradamus's most extraordinarily accurate and detailed prophecies concerned the tragic death of King Henry II of France on July 10, 1559, following a horrendous accident while jousting against the young captain of his Scottish bodyguard, Gabriel Montgomery. The accident happened at what is now called Place des Vosges in Paris on July 1.

A long, deadly splinter from Montgomery's shattered lance pierced the king's golden visor and entered his eye, passing through part of his brain and out through his ear. It took nine days of unspeakable agony for Henry to die. Nostradamus's prophecy, written four years before the tragedy, said:

> Le Lion jeune le vieux surmontera,
> En champ bellique par singulier duelle
> Dans caige d'or les lui crevera
> Deux classes une, puis mourir, mort cruelle.

This translates as:

> The young lion will overcome the old lion,
> In single combat, on a military field,

Place des Vosges in Paris, where King Henry II met with a fatal jousting accident that Nostradamus had predicted.

Inside a golden cage his eyes will be pierced,
Two wounds from one blow, and afterwards a painful death.

An unforgettably poignant prophecy was Nostradamus's forecast of his own death:

He will be able to do no more.
He will have gone to God.
His family, friends and brothers will find him dead
Between his bed and his bench.

And that is exactly how he was found during the night of July 2, 1566. It was a special feast day in honour of Notre Dame, St. Mary the Virgin. Was that, perhaps, rather more than coincidence?

Coinneach Odhar, otherwise known as Kenneth Mackenzie — the Brahan Seer — is harder to pin down historically than either Mother Shipton or Nostradamus. He is thought to have been born near the beginning of the seventeenth cen-tury at Baille-na-Cille, part of the parish of Uig on the Isle of Lewis in the Outer Hebrides, west of the Scottish mainland. His powers of glimpsing future probability tracks were said to come from his use of a seeing stone — one that was either transparent or had a hole in it through which he peered.

These so-called seeing stones feature in several myths and leg-ends associated with prophets and seers. The stories associated with their origins tend to relate how they came into the hands of the prophet as a gift from some para-normal being. This has led some researchers to wonder whether the

The Brahan Seer, whose gifts equalled those of Mother Shipton and Nostradamus.

seeing stones are not legendary at all, but advanced technological equipment. When we think what can be achieved even with our limited terrestrial twenty-first-century technology — digital mobile phones that can pick up e-mails and the Internet, and take and show photographs and video clips — what might superior extraterrestrial technology be able to achieve? The character described as a ghost in the Brahan Seer legend *might* have been an extraterrestrial.

According to the legend of the Brahan Seer, his mother, a fearless Scot, was tending her cattle late at night by the Ridge of Cnoceothail, overlooking the ancient burial ground of Baille-na-Cille, when she saw all the graves opening and a great host of the dead emerging. More curious than frightened, she watched them speeding away in different directions. She stood her ground, continuing to tend her cattle, until an hour later the dead returned and re-entered their graves — which closed silently over them again as though the ground had never been disturbed. Then she noticed that one grave was still open. Wondering why this particular spirit had not returned, she walked to the empty grave and, in accordance with the old folklore and tradition of Lewis, she placed her distaff across it. A distaff is, of course, normally used when spinning flax or wool. It holds the unspun fibres straight, so that they are easier to spin. However, it would not be unusual for a herdswoman to carry her distaff — a sturdy wooden rod — with her when out tending cattle by herself. It could in emergency be used as a weapon, as well as for controlling the animals.

Within a few moments, the spirit of a very beautiful young woman approached the barred grave and asked for the distaff to be removed so that she could re-enter her resting place.

"First, tell me why you were so long after the others in returning," replied the intrepid Mrs. Mackenzie.

"I am a Norwegian princess," replied the spirit in a kind and friendly way. "I was drowned while bathing in the sea near my home, and my body was washed up on the shore here near Baille-na-Cille. Kind and generous folk conducted a funeral service for me — an unknown stranger — and laid my body here to rest in peace until the Day of Resurrection. I took longer to return tonight because I travelled all the way to Norway and back. Now for your bravery, and in gratitude for my funeral, I will give you a gift. Go to yonder lake, and by its edge you will find a round blue stone — as clear as glass — give it your son, Kenneth, and with it he shall see what may come to pass in the future."

Mrs. Mackenzie removed the distaff, and the beautiful Norwegian princess returned to her grave. At the edge of the lake lay the small, transparent blue stone that the princess had promised her, and Mrs. Mackenzie duly retrieved it and gave it to her son, who worked as a labourer for the Seaforths on their Brahan Estate, close to Loch Uissie.

Among his more memorable predictions, the Brahan Seer prophesied that in future ships would sail around the back of Tomnahurich Hill. This was fulfilled when the Caledonian Canal was constructed, largely by the efforts of William Jessop and Thomas Telford. Its locks are spectacular — especially the amazing eight-lock structure at Banavie, often referred to as Neptune's Staircase. Although partially finished in the 1820s, it took a further quarter of a century to reach completion. Just as the Brahan Seer had predicted, it runs around the back of Tomnahurich and joins the Moray Firth at Clachnacudden.

According to another of Coinneach Odhar's strange forecasts, a cow would give birth to a calf in Fairburn Tower. He said that the Mackenzies of Fairburn would lose all that they possessed, and that their once imposing castle would be derelict. This duly happened. In the mid-nineteenth century, a heavily pregnant cow foraging for more hay followed a trail of it up the wide old steps of the ruined Fairburn Tower. At the top, either afraid to descend or because the calf was ready to be born, the cow calved in the upper storey of the tower. Her owner and his men rescued her and the calf and brought them safely down again to their pasture.

The Seer's reference to a one-legged giant from Nigg, who would breathe fire, is thought by some of his supporters to refer to the Ninian Central Oil Platform, which stands on one great pillar and burns off gas periodically. There is an oil rig facility at Nigg, across the Cromarty Firth.

Studying the strangely accurate prophecies of Mother Shipton, Nostradamus, and the Brahan Seer could provide further useful clues to the mysteries and secrets of time. Some of their prophecies are remarkable parallels to events that later took place in our experiential reality. It would be difficult to dismiss their words as mere coincidences. It would be equally difficult to argue with any force that all of their strangely accurate forecasts were mere interpretations by later writers who had already seen and experienced the things that their prophetic heroes were supposed to have foretold from the past. The most objective and logical interpretation of the evidence provided by Mother Shipton, Nostradamus, and the Brahan Seer is that from time to time they became aware of *something* that was no part of their own pragmatic world. They were in touch with something from *elsewhere*. Future probability tracks? The inexplicable Worlds of If? What *was* it that their abnormally sensitive perceptiveness revealed to them?

The legend of the Brahan Seer's transparent blue stone is particularly intriguing. Once again, *water* is integral to the mystery. The stone lay beside the lake. There are echoes of Arthurian legends here: the Lady of the Lake who caught and brandished Excalibur and the lady from the mysterious, enchanted island of Shalott, of whom Tennyson wrote. What was the true identity of the spectral being whom legend describes as the ghostly Norwegian princess who gave the seeing stone to the Brahan Seer's mother?

Other curious examples of apparent prophecy exist. It was forecast that Sir Henry "Hotspur" Percy (1364–1403) would die at Berwick, which he assumed meant the Berwick that was his family home in the north. However, when about to start fighting the

Statue of Henry IV of England, on the tower of Battlefield Church.

Battle of Shrewsbury, he realized that he had left his favourite sword back in the camp — in a nearby village, of which he did not know the name. His faithful squire, however, did know, and offered to fetch Hotspur's sword for him. "It's not far to Berwick, sir," said the lad. Hotspur realized that there was more than one Berwick, and that he was about to fight in the vicinity of the one near Shrewsbury. It was his last battle. An arrow to the head killed him. Another strange prophecy that was fulfilled in a roundabout way, like the Hotspur prediction, concerned the death of Henry IV (1367–1413) "in Jerusalem." He actually died in the Jerusalem Chamber in London.

These examples of prophecy bring the argument round yet again to an ability to perceive future probability tracks not an absolute, rigidly unchangeable future. We never see our destiny because our moment-by-moment choices constantly affect it: we only glimpse what *might* happen *if* ...

Chapter 11

CONQUERERS OF TIME — CASES OF INCREDIBLE LONGEVITY

IN ADDITION to the eternal and abundant life of heaven or paradise that follows life on Earth — the vital truth at the heart of all great world religions — folklore, mythology, and legend make persistent claims of terrestrial longevity. Examples include numerous Rip van Winkle–type stories involving centuries of sleep, the Fountain of Youth, mysterious Shangri-La, and tales of village elders (whose longevity may well be exaggerated by their followers as a mark of respect for seniority). There are political motives for promoting stories of longevity: Stalin, for example, came originally from Georgia and included tales of Georgian longevity in his propaganda. As young teenagers we tend to want the prestigious "street cred" that goes with being nineteen or twenty. From twenty to our mid-thirties, we feel more or less content to play fair with the calendar — but from forty and over there is a widespread desire to look younger. Many of the colourful charlatans running the notorious Wild West medicine shows in nineteenth-century America claimed that their snake oil and tablets with secret ingredients would restore youth and health for just a few dollars. It was part of their spiel to produce an elderly person who appeared to be as fit, healthy, and energetic as someone much younger and then to claim that he was living proof that the patent medicine they were selling was as effective as they said it was. Another angle of the spiel was to exaggerate the age of their "exhibit" and then claim that his long, happy, and active life was entirely due to their medicine.

The desire for finance and fame can also wreak havoc with the truth about people's ages. P.T. Barnum (of Barnum and Bailey fame) exhibited a slave whom he had purchased before the Civil War, claiming that the lady, Joice Heth, was over 160 years old and had once been George Washington's nurse! She was actually only about seventy.

Any study of longevity, its causes, and the light it may shed on the study of time needs to take into account non-human lifespans as well. The bristlecone pine in the White Mountains of California is estimated to be over 4,700 years old — and is still merrily producing cones. Whales have been known to live more than 200 years, and *Lamellibrachia Luymesi*, a deep-sea tube worm, can reach 250. Recent research into sea urchins, dated via radioactive tracings from the Second World War, give them a lifespan of at least 60 years, and some of the researchers involved in the project regard 150 years as a possibility. A giant tortoise reached 255 years, and the oldest known chimpanzee is 74.

One area of the longevity question that attracts a great deal of thought from academic Bible scholars is the longevity of the patriarchs as recorded in Genesis: Adam reached 930, while Methuselah reached 969. Exaggeration of patriarchal ages can be explained as an attempt to forge links between God and humanity. In Japanese history it has been suggested that the ages of the early emperors were adjusted by chroniclers in order to take the dynasties back to a crucial date: 660 B.C.

Myths and legends concerning the Fountain of Youth can be traced back to the trauma of the Black Death plague outbreak in Europe in the first half of the fourteenth century. When the Spanish Conquistadors sailed to the New World, they were looking not only for colonies and gold but for these miraculous fountains that would reverse the ageing process and produce limitless youth, health, and energy. Ponce de Léon was hunting for them in Florida in 1513.

Numerous claims have been made about the extreme longevity of the people of the Hunza Valley in northern Pakistan. The valley itself is 2,500 metres above sea level, and consequently enjoys air that is largely unpolluted. The traditional Hunza diet contains many ingredients derived from local apricots, and some researchers into Hunza longevity have suggested that these are beneficial.

Other claims for extreme longevity seem to be based on religious practices — especially meditation. Taoism, for example, which sets great store by contemplation, was, in its earlier days, concerned with alchemy and the quest for the Elixir of Life. Claims have been made for Taoist longevity extending to as much as two centuries. Swami Bua, a leading exponent of Taoism, claims to have been born in the nineteenth century.

In 1933, it was announced that Li Chung Yun had died in China at the age of 253, having been born in 1680. London's Thomas Parr was buried in Westminster Abbey in 1635 accompanied by the claim that he was over 150 years old. Henry Jenkins was claimed to be almost 170 when he died, while Thomas Carn was said to be 207 when he died in 1588. An amazing old Irish countess was reputedly a sprightly 140 when she died as the result of a fall from a cherry tree while picking fruit for her breakfast. Jeanne Calment was born in 1875 and died in 1997 at the age of 122. Christian Mortensen died the following year at the age of 115.

One of the most persistent traditions of abnormal longevity and time-defying powers is centred on the life of the mysterious Count of St. Germain (sometimes listed as St. Germaine). This remarkable character was thought by some historians to have been the son of Francis II Rákóczi, an exiled Transylvanian prince. Another theory suggests that he was an illegitimate son of the widowed queen of Charles II of Spain, Maria Anna Pfalz-Neuburg. In the eighteenth century, record keeping was far from perfect — it was as easy to steal, change, or assume an identity as it was to lose one. Records of births and deaths were not rigorously certified or maintained. A cunning impostor could have claimed to be anyone at all. No identification papers carried photographs. If it became financially or politically expedient to disappear as character A and reappear in another country as character B, then it was achieved with little or no difficulty. Money was the key that opened the doors of identity and travel, and the man who frequently called himself the Count of St. Germain was never short of money. At one time it seemed that he had the secret of either making diamonds or removing the flaws from them to vastly

Portrait of the enigmatic Count of St. Germain — a man of mystery who seems able to defy time.

increase their value. That was by no means the most significant of the powers that were claimed for him. He was regarded as a fearless adventurer, an intrepid soldier, a profoundly wise statesman, and a chivalrous courtier. He was said to be a scientist far ahead of his time as well as an inventor and musician. With all his powers, he never exploited anyone, never took unfair advantage, never stole, and never cheated. His chief characteristics — apart from his air of mystery — seem to have been his high ethical and moral standards. St. Germain may not have been a saint in the formal, religious sense, but his conduct by human standards was exemplary. If, as many serious researchers have wondered, his origins were far earlier than the eighteenth century, who was he really, and how did he achieve his time-defying longevity? There are theories that St. Germain is still very much alive and active today, playing a prominent role in one or more of the contemporary organizations associated with the ancient Templar-Masonic Guardians of humanity.

The Grand Duke Gian Gastone was a descendant of the very astute Medici dynasty, and there is some evidence that he recognized St. Germain and sponsored his education at Sienna University in Italy. It had an academic reputation that was second to none, and the knowledgeable old Medici may have been a believer in Proverbs 9:9: "Give instruction to a wise man and he will be yet wiser."

There is some evidence that St. Germain turned up in London in 1743 and in Scotland in 1745. Charges of spying were levelled against him there, but after being cleared of these, he acquired his reputation as an outstandingly accomplished violinist. When he met Horace Walpole in 1746, Walpole noted St. Germain's air of mystery as well as commenting on the count's excellent singing, violin playing, and all-round musical brilliance.

The next firm evidence about St. Germain dates from 1758, when he was in Versailles and stayed at the elegant Chateau de Chambord. He had valuable, secret recipes for dyes and pigments, which were in great demand from wealthy aristocrats who spent fortunes on their clothes. During his time there, he met King Louis and Madame de Pompadour (1721–1764), the royal mistress, who, according to some accounts, became St. Germain's mistress as well while he was in Versailles. It was during these fateful years from 1758 to 1760 that St. Germain was said

Château Chambord, where St. Germain was a visitor.

to have bestowed numerous diamond gifts on his Parisian friends — including the exquisite and vivacious Pompadour.

It was also during this period in France that he dropped hints about his time-defying longevity — apparently extending over many centuries. At one soiree an elderly countess peered at him in amazement: "You are the Count of St. Germain," she whispered. "I met you seventy years ago at a ball in Vienna when I was a teenager — and you look exactly the same as you did then!"

"Madame," said the count, smiling and bowing to her gallantly. "I remember you well. You were the most beautiful girl there. It was a privilege to dance with you." Did the Count have the power to *defy* time and remain in his prime for centuries, despite the years that rolled past him? Or did he somehow *step outside* time for years, or even decades or centuries?

A man with St. Germain's many powers and talents attracted jealous enemies. The unpleasant and suspicious duc de Choiseul, then the French Minister of State, tried unsuccessfully to have St. Germain arrested, but the count wisely escaped by heading for England via the Netherlands.

Arriving in Russia at a crucial time, St. Germain is believed to have been in St. Petersburg and to have played a significant part in the army coup that put Catherine the Great on the throne. He then went to

Belgium — known in the eighteenth century as the Southern Netherlands — changed his name to Surmount, and bought substantial land holdings there. While in Belgium, he met and negotiated with a high-ranking Minister of State named Karl Cobenzl, and hinted at his own royal ancestry. Cobenzl was more than a little impressed when St. Germain appeared to be able to transmute iron into gold. For over ten years after his meeting with Cobenzl, St. Germain vanished as effectively as Benjamin Bathurst did at Perleberg. There was simply no sign of him. Does that seem to indicate that his mysterious power over time consisted of an ability to step in and out of it at will, rather than to live through it but be biologically unaffected by it?

He reappeared in Bavaria in 1774 using the title Count Tsarogy and calling himself Freiherr Reinhard Gemmingen-Guttenberg. Two years later he changed his title again to Count Welldone and was seen again in Germany. Once again, he offered not only secret recipes but actual processes that improved wood, leather, bone, ivory, and paper. He also sold wines and liqueurs as well as cosmetics — and yet again he claimed that he possessed the necessary alchemical knowledge to transmute base metal such as iron and lead into gold. His jealous enemies again lied about him, saying that he was preaching blasphemous nonsense and claiming to be either God's special messenger or even God himself. He made no denial of his Freemasonry, and it may have been his Masonic connections that made him the friend and protégée of Prince Karl of Hesse-Kassel, Governor of Schleswig-Holstein. While he was with Karl, St. Germain spent most of his time making herbal remedies, which he gave away to the poor. The name and title that St. Germain used while he was with Karl was Francis Rákóczi II, Prince of Transylvania.

The friendship and mutual trust between St. Germain and Karl went deep. It is generally conjectured that the report of St. Germain's "death" in 1784 was a convenient fiction that the two of them had constructed. St. Germain turned up again very much alive and well — and still showing no perceptible signs of age — in Paris in 1835. He surfaced again in Milan in 1867, and Napoleon III (the "Little Napoleon") kept a careful dossier on him. The famous Annie Wood Besant (1847–1933) claimed that she had met the count in 1896. C.W. Leadbeater, a pioneer of theosophy, reported that he had met St. Germain in 1926 in Rome. During their encounter, Leadbeater said that the count had shown him a very interesting ancient robe, which he claimed had once been worn by a Roman emperor. St. Germain also told the theosophist that he owned and occasionally lived in a castle in Transylvania. As recently as 1972, a man on French television claimed to be St. Germain: he could have been an impostor, or simply a publicity seeker — but the possibility that he really *was* St. Germain cannot be dismissed out of hand.

Over and above the basic facts that make up most of what is known of St. Germain's mysterious life, many claims are made about him that may be pure fabrications and exaggerations. The Count of St. Germain was seen from different perspectives by different groups. Theosophists regarded him as an adept, a master, or a mahatma. Aleister Crowley thought of the possibility of passing himself off as St. Germain — or even of attempting to become a "type" of St. Germain, and may have led some of his more credulous followers to think that he actually *was* St. Germain. The mystical and magical Madam Helena Blavatsky claimed that St. Germain was one of the Masters of Wisdom from whom she had learned much of her secret lore.She also claimed that St. Germain had access to, or was in possession of, various secret documents containing vitally important information. The first thought that occurs to researchers who are familiar with the destruction of the Cathar fortifications at Montségur in southwestern France is the enigma of the four fearless Cathar mountaineers who descended the precipitous crag on which Montségur stands, carrying with them "the treasures of their faith." According to Latin

The famous Annie Besant, who reported that she had met the mysterious Count of St. Germain.

depositions of the Holy Inquisition, these treasures amounted to pecuniam infinitam (literally "unlimited money"). That in itself was a strange choice of phrase. A lot of money, a great deal of money, a vast fortune ... any such description of the Cathar treasure (whatever it was) would seem more natural and rational than to call it infinite money. The concept of infinite money makes sense only if the secret was some technique for creating money — something like the fabled Philosopher's Stone that turned base metals into gold.

As mentioned in an earlier chapter, co-author Lionel interviewed Dr. Guirdham (an acknowledged world authority on the Cathars) on the subject of the priceless items that the four intrepid Cathar mountaineers took with them on that perilous descent. Dr. Guirdham had a quick, certain and precise answer: "Books, my boy, books. They were carrying books." If we now imagine that the ageless St. Germain — by that name or some other — was in the vicinity when the Cathar stronghold of Montségur fell to the army of Catholic Christians in 1244, we may also imagine those vitally important secrets passing from the doomed Cathar garrison into the sacred, safe custody of Count St. Germain.

With Helena Blavatsky's deep knowledge of mysticism and magic, did she know, or guess, that the secret documents in St. Germain's care were from the Cathars of Montségur?

There are other researchers who take the mystery of St. Germain's secret documents much further back into the mists of time than the massacre at Montségur in 1244. The Count of St. Germain was by no means the only timeless mystery man. Hermes Trismegistus (or "thrice-blessed Hermes") was also alleged to be one and the same

Madam Blavatsky claimed that she had learned important secrets from St. Germain.

entity as the immortal Thoth, scribe of the gods of Egypt. In an old Hebrew legend, Thoth was asleep, hibernating, or in a state of suspended animation in a cave when Abraham's sister-wife, Sarah, entered the cave and found him there. She was understandably captivated by the gleaming emerald tablets that lay close to his motionless body. According to legend, it was upon these mysterious emerald tablets that the secrets of power from Atlantis (and the Atlantean colony of Egypt) were inscribed. In these inscriptions lay all the power and wisdom of the Egyptian gods. Sarah picked up two of the stones and was turning them over in her hands and admiring them when the recumbent Thoth began to stir. Sarah fled from the cave in terror — still clutching the two priceless tablets.

Developments of the ancient Hebrew legend suggest that these awesome emeralds became the Urim and Thummim of the Old Testament, the mysterious divinatory stones that certain high priests who knew their secret used. Tragically, the Urim and Thummim were lost — perhaps at the time of the Jewish Captivity in Babylon — and the Old Testament records that knowledge of their use was also lost. Perhaps the last high priest who understood them properly died before he could pass on their enigmatic secrets.

There is also a viable possibility that such important and valuable knowledge would have been secretly committed to writing. Was *that* what was written on the strange documents that Helena Blavatsky maintained were in the possession of the Count of St. Germain?

Other ideas about St. Germain identify him with Cartophilus, the Wandering Jew of legend, and with the origin of Rosicrucianism. Gian Giacomo Girolamo Casanova (1725–1798) claimed that he had met St. Germain, which adds another dimension to the mystery. Most accounts of St. Germain describe him as celibate, spiritual, and ascetic: the precise opposite of Casanova. Yet there is the persistent story of St. Germain's affair with Pompadour, and in his courteous treatment of the elderly countess who recognized him from seventy years before there is a hint that on their first meeting they had done rather more than dance together.

St. Germain is regarded by numerous New Age groups as having the rank of an ascended master, and many of them credit him with the ability to use the whole array of psi powers: telepathy, teleportation, telekinesis, levitation, and the ability to glide through solid matter as if he were a ghost. His reputed mastery of and transcendence over time seem to fall into that psi category as well.

Advocates of chromotherapy and believers in the power and influence of colours always associate St. Germain with purple and violet. He is sometimes given the title of Master of the Seventh (Violet) Ray and is also regarded by some groups as the Master of the Age of Aquarius. Alice Bailey's books about St. Germain give him the title of Lord of Civilization and give details of his prophecy to the effect that after 2025 he will reappear physically on Earth along with other supreme spiritual masters. Other writers on St. Germain have identified him with Joseph (husband of the Virgin Mary), Arthur's magician Merlin, Roger Bacon, Francis Bacon, and Christian Rosenkreuz. He is also credited with being a high priest among the white magicians of Atlantis, possessing the secret of great longevity and many other amazing powers.

Those St. Germain theorists who regard him as Francis Bacon suggest that although Bacon allegedly died in 1626, someone else was buried in his coffin, and he himself (really St. Germain) carried on with the work he had been already been doing for centuries via his many changing identities. The most recent St. Germain conspiracy theories envisage him as being alive and well and secretly advising the American government — along with a number of benign extraterrestrials.

One of the greatest mysteries associated with the Count of St. Germain is his reputed mastery of time: If he did conquer time, how did he accomplish it? Was he able to side-step it and watch it rolling harmlessly past him without affecting him, or could he drive through it at different speeds when the fancy took him, like an expert rally driver mastering difficult road surfaces or cross-country terrain? Was his conquest of time simply inside his immensely powerful mind? The power of mind over matter is a far greater power than is generally realized. Could St. Germain extend his mental powers objectively as well as subjectively? Did he simply tell himself not to succumb to the ageing process, or could he control time itself by using the power of thought? Whatever he did, it seems to have been a talent that he was able to pass on, because his faithful valet — the equally enigmatic Roger — seems to have been able to do it too. Further speculation suggests that Roger was really an alias for Cagliostro, and the St. Germain mystery grows and intertwines with other enigmas of the period. Franz Graffer and others claimed that St. Germain could read minds, control reptiles, write with great speed and efficiency using both hands, and play an entire string quartet's pieces by himself, using only one violin. Supposing that this last claim contains some vestige of truth — a man with a mysterious mastery of time would be capable of stepping momentarily forwards or backwards in it to play the harmonious accompaniment to his own violin melody, and in so doing would produce the same sounds as a whole group of musicians playing simultaneously.

Count Cagliostro, one of the disciples of the Count of St. Germain — and almost as mysterious as his master.

Another mystery of longevity, and one that perhaps suggests the wider and more complex mystery of parallel lives existing at different times, is centred on the lives of Aulus Cornelius Celsus and Theophrastus Philippus Aureolus Bombastus von Hohenheim, who was known as Paracelsus ("alongside, or the equal of, Celsus"). Aulus Celsus flourished during the closing decade of the first century B.C and the first half of the first century A.D. Although remembered mainly as a physician and surgeon because of his book *De Medicina*, dealing with surgery, diet, and various medicines and healing tinctures, including opioids, Celsus was interested in almost every branch of knowledge. His lost writings covered military skill and tactics, the techniques of rhetoric and argument, and agricultural methods. He was a great believer in the empirical approach to medicine: had he lived today, he would have been an advocate of cloning and genetic engineering. But his approach, as expressed in his introduction to *De Medicina*, was a carefully balanced and objective one. He had strong, progressive ideas of his own, but was always prepared to listen to and consider other people's arguments. He was a man who weighed up the pros and cons before taking action. His knowledge of surgery was far ahead of his era: he wrote about the removal of cataracts and bladder stones, and the treatment of fractures. His work was rediscovered by Pope Nicholas V, who was born in 1397 and reigned in the Vatican from 1447 until his death in 1455. It is particularly interesting to note that the Pope who rescued Celsus's medical wisdom was himself the son of a doctor. As a result of Nicholas's interest and support, Celsus's medical work was eventually republished in 1478 — more than twenty years after the Pope's death.

Theophrastus Philippus Aureolus Bombastus von Hohenheim — known as Paracelsus — lived from 1493 until only 1541, as far as the official records go. But it was by no means unknown at that time for a convenient change of identity to be made when a supposed death occurred.

Paracelsus was born in Switzerland, and as a young man worked in the local mines as an analyst. He began to study medicine at the age of sixteen, gaining his doctorate from the University of Ferrara. Most significantly, he continued his studies in Egypt, Arabia, Constantinople, and Palestine, where he took particular interest in alchemy. He learned many techniques that were far in advance of European medical methods — and, like his famous Roman namesake, he was a man who was centuries ahead of his time. Is it possible that both Celsus and Paracelsus had access to medical technology that did not belong to first-century Rome or to sixteenth-century Europe? Is it even remotely possible that in the course of acquiring that anachronistic information they encountered each other at some mysterious crossroads in time?

Paracelsus, a man who may have lived one life in Rome and another centuries later in Europe.

Among the strange things that Paracelsus investigated were the Pythagorean mysteries, and these were thought by some researchers to hold the secret mathematical clues to controlling time itself. Paracelsus's interest in magic and his quasi-scientific study of those areas of magic that seemed to him to work — albeit spasmodically — led him to write: "Resolute imagination is the beginning of all magical operations. Because men do not perfectly believe and imagine, the result is that arts are uncertain when they might be wholly certain." Was Paracelsus hinting that control over time depended upon the time wizard's ability to *believe* that such control was really possible? Did he think that, provided the time traveller could imagine in every detail what time travel would be like, it could be achieved?

More than thirty years of careful, on-site research into many aspects of the paranormal have convinced the authors that the power of the human mind is close to being limitless — if only we can find and use the keys that unlock its awesome powers. Paracelsus maintained that two of those vital keys were *imagination* and *faith*. In the New Testament records, Christ himself repeatedly teaches the central importance of faith as the mainspring of healing, and of other miracles. The brilliant Felix Dennis, writing in his superb volume *How to get Rich*, says, "Without self-belief nothing can be accomplished."

The authors would argue strongly that faith, self-confidence, determination, and an absolute refusal to admit defeat or accept failure under any circumstances are the maps and compasses that need to be used during the quest for those elusive keys to mind-power.

Paracelsus was also deeply involved in the teachings of Hermes Trismegistus, and may even have had access to some of the ancient riddles said to have been engraved on the Emerald Tablets. Paracelsus may also be included among the earliest dieticians and pharmacists because of his keen interest in the value of certain essential minerals in the human body. Did his advanced knowledge of beneficial medical chemicals come from elsewhere?

He expressed his priorities by saying: "Many have said of Alchemy, that it is for the making of gold and silver. For me such is not the aim, but to consider only what virtue and power may lie in medicines."

Another legendary candidate for abnormal longevity is the Wandering Jew, variously named as Ahasuerus, Ahasverus, Cartaphilus, Cartophilus, and Malchus. In one legend, he made cruel and offensive comments as Jesus was led away to be crucified and was told that he would walk the Earth until Judgment Day and the end of time. In a totally different legend, he is one of the disciples, John, because of a literal interpretation of Matthew 16:28: "some standing here shall not taste of death till they see the Son of Man coming in his Kingdom ..." In the first version of the legend, longevity is seen as a punishment. In the second version it is a reward. There is also the idea that Cain, the first murderer, has to wander the Earth until the end of the world, and there is an Islamic account in the Koran that Moses cursed Sameri for helping to make the golden calf. As a result, Sameri also has to wander the Earth until Judgement Day. In the legend of the Wandering Jew, folklorists suspect that these various accounts have probably coalesced.

In one version, perhaps the best known, Cartaphilus was a Roman gatekeeper who struck Jesus, asked him why he was loitering, and told him to hurry up. Jesus turned and said, "You will wait on Earth until I return." This version is found in Roger of Wendover's book *Flores Historiarum*, which dates from 1228. In a later version, produced in Europe during the sixteenth century, the Wandering Jew is a shoemaker named Ahasverus. As the exhausted Jesus passed his premises on the way to Calvary and tried to rest for a moment, the shoemaker angrily drove him on.

Geoffrey Chaucer's (1343–1400) *Pardoner's Tale* relates the fate of three rash young men who try to kill Death. On the way, they meet a strange old man who tells them where Death is to be found. Chaucer does not name him as such, but there is every possibility that he is intended to be the Wandering Jew — and Chaucer would almost certainly have been familiar with Roger of Wendover's work.

If, for the sake of discussion, we separate the possible existence of such a time-resistant wanderer from all of its mythical, legendary, or religious connotations, the question is: *how* did he resist time? Did the wanderer naturally have a far longer lifespan than the average, terrestrial human being? If so, was he from another planet or another dimension? Was he not so much a being with great longevity as a being who could traverse time and probability tracks?

Great longevity, perhaps even immortality, is an attribute of Melchizedek — who *may* be one and the same as Hermes Trismegistus, alias the Egyptian Thoth, scribe of the gods. He is said to have been the

Priest-King of Salem — almost certainly Jerusalem — and to have met Abraham. Melchizedek's name may have been derived from "Zedek is my king" or from "My king is righteous." It is possible that Zedek was either an historical Middle Eastern king or a Middle Eastern deity. References to "the Most High God" in association with Mechizedek's name make him out to be a priest of Yahweh, or perhaps of some other deity known and worshipped in Salem in the remote past. The critical point about him is that he is regarded as having no parents and "no beginning of days nor end of life ..."

Ancient Rabbinical ideas about Melchizedek associate him with Adam, and credit him with passing on Adam's robes to Abraham. He is also seen as the prototype of another form and order of priesthood, differing from the Aaronic priesthood. The basis of the Melchizedekian Order of priests is eternal life, and Christ himself is referred to in the Epistle to the Hebrews 5:10 as being "a Priest after the Order of Melchizedek."

Whether Melchizedek was simply a Middle Eastern priest-king or an alias of the mysterious Thoth, his most notable characteristic was his apparently unending lifespan. The reference to his having no beginning, no earthly parentage, and no traditional genealogy at a time when genealogy was vitally important hints strongly at the possibility that he came from *elsewhere*. Was he, too, a time traveller or an extraterrestrial visitor? Did Melchizedek move between the dimensions and traverse numerous probability tracks? Once again, the records of him raise the major question about the nature of time. Is it something that can be travelled through? Can we enter or leave it at will, once we have the secret of opening and closing its mysterious portals?

An understanding of the mysteries of credible and incredible degrees of longevity will go at least part of the way towards solving the mysteries of time itself — and human involvement with it.

Chapter 12

THE VERSAILLES TIME SLIP ADVENTURE OF ANNE MOBERLY AND ELEANOR JOURDAIN

IN NOVEMBER 2006, the authors made a special research visit to Versailles to study the site first-hand and to take essential photographs of the area where Charlotte Anne Elizabeth Moberly and her friend Eleanor Frances Jourdain experienced what they firmly believed was a time slip adventure. Moberly and Jourdain wrote everything up in considerable detail, and their research into their strange adventure covered a period of some ten years during which they made repeated visits to Versailles. Writing under the pseudonyms of Miss Francis Lamont (Jourdain) and Miss Elizabeth Morison (Moberly), they produced an accurate account of their strange experiences, which was published as *An Adventure* by Macmillan of London in 1911.

The publisher's note among the preliminaries reads: "The signatures appended to the Preface are the only fictitious words in the book."

The magnificence of Versailles.

The Publishers guarantee that the Authors have put down what happened to them as faithfully and accurately as was in their power."

The accuracy and validity of the account that the two women gave is supported critically and objectively by a hard-headed contemporary scientist, Professor Sir W.F. Barrett, F.R.S. He read their independent accounts of their curious experiences at Versailles, and they gave him access to letters from various friends to whom they had described their strange adventures at Versailles. Barrett said that following his perusal of the documents "no doubt whatsoever" was left in his mind that "the story was written substantially as it appears in this volume." Barrett was not prepared to accept that what had happened to the two ladies had been a genuine time slip. He did accept, however, that the authors "had experienced a remarkable collective hallucination." Charles Fort would

Miss Jourdain, one of the two English schoolteachers who reported experiencing what seemed like a time slip at Versailles.	Miss Moberly, the other English schoolteacher who reported the strange experience at Versailles.

have smiled at Barrett's description of the women's experience in Versailles as "a collective hallucination." In Fort's approach to the mysterious, what people tend to cling to for comfort and reassurance under the umbrella term *reality* may itself be only a collective hallucination.

Our own research visit to Versailles in November 2006 enabled us to follow in Moberly and Jourdain's footsteps and to absorb the atmosphere of these strange old buildings and the gardens and woodlands that surround them. Even with its twenty-first-century signposting and good, accurate maps and plans of the Versailles estate, it is not an easy place to navigate. As we have noticed repeatedly during our research and analysis of these time mysteries, water seems to play a significant role in many of them, and Versailles was no exception.

The ladies began their August 1901 trip by describing how "a very sweet air" was blowing as they sat in the Salle des Glaces — the famous Hall of Mirrors. They walked from there until they reached the end of the Grand Canal that was nearer to the Château, and which the plan of Versailles describes as the Embarcadere. They turned right from the Grand Canal along one of the many woodland paths through Bosquets

Co-author Lionel, standing above the Embarcardere.

One of the woodland paths near the Petit Trianon at Versailles where
Miss Moberly and Miss Jourdain lost their way.

The Grand Trianon at Versailles, which Moberly and Jourdain visited.

du Trianon. Their comment on that day's weather is significant at this point in their narrative. The weather had been very hot, bright, and sunny up until their visit to Versailles, when it clouded over: "the sky was a little overcast and the sun shaded ..." They then reached the Petit Canal close to the Bassin du Fer-a-Cheval, adjacent to the Grand Trianon. Finding the meaning of the term *Trianon* seemed relevant, but it did not appear as such in any French dictionary or Internet translation service to which we had access. However, further research seemed to indicate that the name was chosen because a village named Trianon had been bought by Louis XIV so that he could construct "a house for partaking of collations": a quiet, informal place to which he could retreat with his family occasionally to get away from the demanding protocol of the court. It seems that even the redoubtable Sun King needed a little tranquility now and again!

Miss Moberly and Miss Jourdain correctly identified the Grand Trianon and then did their best to get their bearings from it. Passing it on

Co-author Patricia researching the Domaine de Marie-Antoinette at Versailles.

their left, they came to what is known today as Allée des Deux Trianons, which they described as "a broad green drive, perfectly deserted." Not realizing that it would have led them directly to the Petit Trianon, they crossed it, then turned right and saw some buildings. They were now well and truly inside the Domaine de Marie-Antoinette. Miss Moberly was rather surprised that her friend failed to ask for directions from a woman whom Miss Moberly saw shaking a white cloth out of an upstairs window. Seeing three paths immediately ahead, they noticed that there were two men on the central one and assumed at first they were gardeners because of a wheelbarrow and a spade nearby. They changed their opinion later when they realized that the men were wearing small tricorne hats and long, grey-green coats. This made the ladies decide that they were officials of some sort — perhaps garden supervisors or planners. The men directed the ladies to go straight on along this central path. Miss Moberly then records that an unaccountable feeling of great sadness and depression came over her. She did her best to overcome it in order not to spoil the trip for her friend. As she struggled with these strange, melancholy feelings, they crossed another of the many paths that filled the wood. The next thing they saw was what Miss Moberly described as "a light garden kiosk" with a strange, sinister-looking man sitting beside it. He wore a cloak and a wide-brimmed hat, and she described his face as repulsive and his complexion as rough. Her feeling of depression was now tinged with fear. She said later that things looked strange and unnatural. The trees behind the little wooden structure no longer seemed real. She described them as having the appearance of trees in a tapestry. She asked Miss Jourdain which way they ought to go, but was quite determined to go no closer to the dangerous-looking man near the kiosk.

The ladies' report of the strange arrival of another man at that moment adds further mystery to their account. He seemed to have emerged from the nearby rock, and he was agitated and breathless as though he had been running fast to reach them in order to prevent some harm or danger to them — and they had, in fact, *heard* sounds of running just before he appeared. In stark contrast to the unpleasant-looking character near the kiosk, this newcomer was handsome. He wore a sombrero-style hat over his long, crisp, curly black hair, and Miss Moberly commented particularly on his large, attractive dark eyes. When she looked at him intently, as though trying to work out where he had come from, he gave her what she later described as "a most peculiar smile." Like the man by the kiosk, their rescuer also wore a cloak. He urged them vehemently to take the path straight ahead, and it seemed to the ladies that he was most anxious that they should not turn left past the evil-looking man by the kiosk who had so unnerved Miss Moberly. As she herself had no intention of going to the left, she went straight ahead where their rescuer had directed them. Turning to thank him, she was very surprised to find that he was no longer in sight, although she could again hear the running sounds that she had heard immediately prior to his appearance.

The ladies then followed the path that their rescuer had indicated and found that it went over a small bridge crossing a miniature ravine.

Once again water appears to be an important aspect of time slip mysteries. They reported a fine cascade descending into a tiny stream at the base of the ravine. The ladies also described ferns, but they could, perhaps, have meant rushes, and the authors certainly encountered a similar spot beside a small bridge, with a narrow path running behind it.

Co-author Lionel among the reeds and bulrushes in what may have been the strange brook Miss Moberly and Miss Jourdain described.

They proceeded through part of the forest and by the side of a meadow filled with long grass that gave them the impression of dampness and shadow. Trees obscured their view of the Petit Trianon until they were very close to the building itself. Miss Moberly described the building as smaller than she had expected, and commented on its square solidity. She noticed a terrace running around the north and west sides of the house, and a woman sitting there and sketching. Miss Moberly guessed that the woman was sketching trees, as there was nothing much else in front of her. The woman looked directly at Miss Moberly as she passed, and Miss Moberly saw her face clearly. She commented that the face — although quite pretty — was not a young face, and she did not feel attracted to the woman. If it was indeed Marie Antoinette whom the English visitor saw, it seems a little strange that Miss Moberly regarded her as "not young." Marie Antoinette was born in 1755 and was only thirty-four in 1789, when the Paris mob attacked Versailles. Perhaps the pre-revolutionary political strain and anxiety had aged her prematurely. As history proved, the ferocity of the Paris mob was a formidable thing.

In her account, Miss Moberly noted that the woman whom she saw sketching close to the Petit Trianon was wearing a shady white hat over fair hair fluffed around her forehead. With a keen eye for style and fashion,

The Petit Trianon, which Miss Moberly and Miss Jourdain believed was haunted.

Miss Moberly commented on the woman's light summer dress, arranged on her shoulders in what she called "handkerchief fashion." She also noticed particularly that there was "a little line of either green or gold near the edge of the handkerchief," which indicated that it was over the top of her low-cut bodice, not tucked inside it. The skirt was full and long-waisted, and, by Miss Moberly's rather prudish Victorian standards, "short."

It is more than a little surprising that when the ladies compared notes about what they had seen and experienced that August afternoon at the Petit Trianon, Miss Jourdain had not seen the woman whom Miss Moberly had seen sketching — although Miss Jourdain said that she had felt as if there were people there whom she could not see. Of particular interest was Miss Jourdain's feeling of sadness and depression, which coincided with what her friend had also felt at the point where they had both seen the men whom they assumed to be gardeners.

As they talked through their experiences together, they wondered why the mysterious running man, who had seemingly rescued them, was wearing a cloak wrapped around him on such a warm August day. It even crossed their minds that he was about to fight a duel with the

Co-author Patricia at the Place de Vosges, site of the Nostradamus
prophecy concerning the death of Henry II.

odious-looking man beside the kiosk and wanted to get them out of the way so that he could begin it without witnesses being present.

A few weeks after their adventure in Versailles, a friend of Miss Jourdain's told her that there was a persistent tradition in France that on a particular day in August (most researchers maintain that it is August 10) the ghosts of Marie Antoinette and several of her contemporaries are seen and felt in the area near the Petit Trianon.

She communicated this news to Miss Moberly, who did some additional historical research and discovered that on August 10, 1792, the Tuileries were sacked by the Paris mob. The Tuileries themselves had an interesting history. As described in Chapter 10, one of Nostradamus's most detailed and significant prophecies concerned the tragic jousting accident that led to the eye injury and consequent agonizing death of Henri II in 1559. This jousting accident took place at what is now known as the Place des Vosges, opening on to the Louis XIII Square.

Henri's widow, Catherine de Medici, arranged for the Tuileries to be built close to the Louvre. The name is derived from the tile kilns (French *tuileries*) that once stood on the site. Louis XIV, the Sun King, lived in the Tuileries while Versailles was being built. The hapless Louis XVI and Marie Antoinette and their family were forced to leave Versailles and were placed under house arrest in the Tuileries. On August 10, 1792, the Tuileries were stormed by the Paris Mob, who massacred the Swiss Guards, while Louis XVI and his family ran for their lives through the gardens to the Hall of the Legislative Assembly, where they took refuge.

Miss Moberly came up with an interesting theory to the effect that on that fateful August 10, 1792, Marie Antoinette might have been thinking deeply about happier days in Versailles, close to the Petit Trianon that had meant so much to

The Louvre.

her. Miss Moberly wondered whether she and Miss Jourdain had inadvertently walked into some strange projection of the doomed queen's vivid memory, and that that was what had accounted for their feelings of deep sadness while they were near the Petit Trianon on their August 1901 visit.

The idea of being caught inside someone else's dream is one that has been handled skilfully by a number of fantasy writers, and L. Ron Hubbard's *Typewriter in the Sky*, first published in serial form in 1940, has the hero trapped inside a story that one of his friends is writing. The Reverend Charles Lutwidge Dodgson (1832–1898), better known as Lewis Carroll, was a philosopher, logician, and mathematician as well as an author, and *Alice's Adventures in Wonderland* had his young heroine involved in a brilliantly imaginative dream adventure. With Dodgson's academic powers, his second volume — in which Alice passes through a mirror into a back-to-front, future-to-past world — contains several extremely deep and serious ideas that touch on the nature and reversibility of time. It is by no means impossible that Miss Moberly's theory of being involved in a projection of Marie Antoinette's thoughts owed something to her reading of Carroll's works.

Miss Jourdain's second visit to Versailles in January 1902, took place on a cold, wet day, and because time was of the essence for her on that occasion, she went straight to both Trianons, also inspecting the Temple de l'Amour as part of her research. This inspection satisfied her beyond any reasonable doubt that it was definitely not the same place as the strange kiosk beside which they had seen the sinister man on their August visit. Up to this point in her January 1902 visit, she felt none of the unnatural fear and sadness — and the weird sense of unreality — that both ladies had experienced the previous summer.

Then she crossed the bridge leading to the Hameau. In her own words, she felt as if she had "crossed a line," and the unwelcome feeling of strangeness, sadness, and fear returned in full force.

She noticed two men loading sticks into a cart. Both were dressed in the characteristic tricorne hats and cloaks that the two ladies had seen on their August trip. But instead of the grey-green colours that they had noticed then, one of these men was wearing dark blue, and the other had a brownish red cape. Miss Jourdain felt relieved to see them and thought that if she needed help with directions, she could ask them.

For scarcely more than a moment, she turned to look again at the quaint architecture of le Petit Hameau, also known as le Hameau de la Reine ("the little hamlet belonging to the queen"). When she turned back, the men and their cart were nowhere to be seen — and there were no signs of any sticks on the ground at the place where the men had been loading them. With a great deal of courage and determination, she continued towards the Hameau, and once again, water features prominently in the time slip equation. As the authors noted on their research visit, the buildings in the Hameau are all on the edge of a landscaped pond. The architect responsible for it was Ange-Jacques Gabriel (1698–1782), one of the greatest French architects of the eighteenth century. Louis XV had it built for his mistress, the beautiful and brilliantly intelligent Madame de Pompadour. Louis XVI gave it to Marie Antoinette.

The Hameau was what was known in France at the time as a *ferme ornée*, an ersatz or artificial farm where wealthy aristocrats could play idyllic, romantic games dressed as milkmaids and shepherdesses. Handpicked and extra docile cows were washed and groomed there as an integral part of this fantasy farming. There were even monogrammed ceramic milk churns, made in the royal porcelain works at

The Hameau Village, a ferme ornée, part of Marie-Antoinette's Domaine, where Miss Jourdain experienced strange psychic feelings. Once again, water seems associated with strange time slip reports.

Sèvres, in which the genteel milkmaids would store the milk the placid cows produced!

As Miss Jourdain walked from the Hameau through dense trees, she became aware, just as she had done before, that there were invisible people walking close to her — so close, in fact, that she could hear what they were saying to one another.

Another factor that struck the ladies as they later pondered over their strange adventure was that during their 1904 visit, the grounds

Artist's impression of the painting of Marie Antoinette and her children by Swedish artist Adolph Ulrich Wertmuller that resembled the woman whom Miss Moberly saw at Versailles.

and buildings were crowded with tourists (as they were when the authors went there in 2006), whereas in 1901 Miss Moberly and Miss Jourdain had seen scarcely anyone. Was it a private royal precinct from the past that they had inadvertently stepped into? Or were they — in some incomprehensible way — wandering inside the powerfully projected thoughts and memories of the doomed Marie Antoinette?

Understandably, there was a great deal of controversy over the mysterious account that the two ladies gave, but they sturdily defended their version of the strange events and their interpretation of what they had seen and heard.

Anne Moberly later saw a picture of Marie Antoinette and her children painted by the Swedish artist Adolph Ulrich Wertmuller (1751–1811) and recognized her as the woman she had seen sketching near the Petit Trianon. They had also noticed a plough while they were walking through the grounds in 1901 — but no plough was there in 1901, although there had been one in the past. They never found the elusive bridge nor the strange kiosk on later visits, but they did find an old map indicating that there had been a gazebo there at one time, which had been demolished before 1901. They were more or less able to identify the sinister-looking man with the pockmarked face as the evil Comte de Vaudreuil, who had treacherously helped to bring about Marie Antoinette's downfall. Their research also revealed that the men whom they had seen in the tricorne hats were no part of the Versailles of 1901 — they were actually wearing Swiss Guards' uniform from the late eighteenth century. The mysterious breathless runner who had insisted that they not go to the left could well have been a loyal servant of the royal family sent to warn Marie Antoinette that the Paris mob was approaching.

So what are the possible explanations of the Moberly and Jourdain time slip experiences at Versailles, and what conclusions can be reached from their evidence? After both ladies had died, their real identities were revealed, and their academic status gave increased credibility to their evidence. If it was a straightforward time slip that had taken them back from 1901 to 1789, that would account for the absence of visitors to the royal domain. The curious detachment of some of the people they saw — and the ways in which they either vanished rapidly or were not always visible to both ladies — might suggest that whatever glitch or fault had affected time that August day had irregular and inconsistent qualities, as though the edges of time had been roughly torn.

As Paracelsus taught, the power of the mind is vast, and emotion, such as Marie Antoinette must have felt in those terrifying days, *could* have been capable of creating a strange quasi-realm into which Miss Moberly and Miss Jourdain were drawn.

Chapter 13

ANACHRONISTIC ANOMALIES
AND ARTIFACTS

WHEN PEOPLE, animals, or objects turn up where they aren't supposed to be or when they aren't supposed to be, it raises all kinds of
intriguing questions. The last coelacanth, for example, was supposed to
have gone to a better world around 10 million years ago — then his clan
turned up happily swimming around off the coast of Africa! What
seemed very similar to part of a modern-looking shoe was found in
Triassic limestone in Pershing County, Nevada, United States, by an
archaeologist called Alfred E. Knapp in 1927 — or by a different archaeologist named John T. Reid in 1917. The two-version account presents
something of a challenge to researchers! What looked like a crystal magnifying glass turned up at Nineveh, and gold thread emerged from a quarry in Scotland. Greek letters appeared inside ancient marble, and a strange
steel cuboid was found in an Austrian mine. Open-minded, objective
examinations of the evidence would suggest that these anachronistic discoveries might signify visits from intelligent extraterrestrials long, long
ago — or might indicate that some sort of time travel had taken place.
The authors approach these reports of anachronistic artifacts with their
own brand of Fortean philosophy: "Nothing is so seemingly ridiculous
that it isn't worth investigating; nothing is so conclusively and definitively proved that it is beyond any further investigating."

Significant controversies arise over the numerous reports of
anachronistic anomalies and artifacts that have been found over the
years. The fundamental source of these controversies is that those who
believe devoutly in Creationism want the anachronisms to exist because
they will cast doubt on Darwin's theory of evolution. The supporters of
evolution would prefer the anachronisms to be revealed as deliberate
hoaxes, honest misinterpretations in good faith, or simple misunderstandings. They do not then have to try to explain how certain examples

of advanced technology came to be in rock strata that were thought to be millions of years old. Some enthusiastic Creationists are, therefore, tempted to accept anachronisms rather less critically than keen Evolutionists do. In that kind of ideological battle, objective, scientific truth can all too easily become a casualty.

The types of arguments, claims, and counter-claims that swirl around what are said to be anachronistic objects are clearly illustrated by the Nevada Heel incident. Something odd was found in Fisher Canyon, Nevada. It could have been a rather unusual but perfectly natural and explicable rock formation that just happened to resemble a modern shoe. Or it could have been part of a shoe that a time traveller had left behind. A third possibility would be that a humanoid, extraterrestrial visitor had left it here. A further idea would involve an entity from another probability track, or even from a different dimension. The more objectively, intensely, and accurately these anachronisms are studied, the more we shall learn about the mysteries and secrets of time. Here are a few of the more dramatic and interesting cases that are well worth re-examining.

The so-called Babylonian electric cells were found by Austrian archaeologist Wilhelm Konig in 1931. He later became director of the Baghdad Antiquities Administration, working from the Iraq Museum.

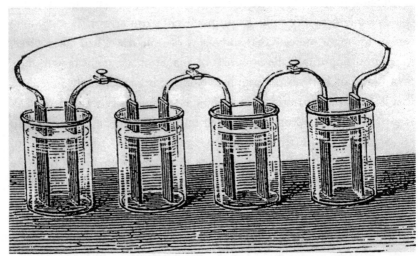

Galvanic cells of a type that would have been familiar to Konig in the early twentieth century, when he examined the strange, ancient Parthian artifacts.

Digging at a Parthian site in Khujut Rubu'a, he came across a small ceramic container with a copper cylinder inside it. This had been soldered with an alloy of tin and lead, topped by a copper disc, and sealed with bitumen. An iron rod showing acidic damage was secured within the copper cylinder. In Konig's opinion, the only possible explanation of the artifact was that it was an electric cell, and his theory was justified when working reproductions of it produced a potential difference of about one volt. Konig's example from Khujut was by no means unique: numerous other examples were found in the region — all dating from the Parthian period between 300 B.C. and 300 A.D. A significant part of the mystery is *why* the Parthians were using electric cells over two thousand years ago. No devices have yet been found that Konig's cells might have powered.

In 1900, a Greek diver looking for sponges near the tiny island of Antikythera came across an ancient wreck containing a veritable Aladdin's cave of interesting objects in its cargo holds. Most of them found their way to the National Museum in Athens, where one that looked like a lump of rotten wood and corroded metal lay more or less unnoticed for years. It was a historian of science, Derek de Solla Price (1922–1983), who began examining it using gamma-photography. His painstaking research revealed intricate gears that enabled the device to show the relative positions of the sun and planets: the sponge diver had salvaged an ancient astrolabe of remarkably fine workmanship. Its delicate gearing was based on the ancient Ptolemaic system, with Earth at the centre of the solar system and the sun, moon, and planets revolving around it. Some researchers have described it as a computer rather than an astrolabe, but astrolabe seems to be a more accurate term for it.

In 1856, a tunnel was being constructed to carry French railway lines from Nancy to St. Dizier. It was reported that as the workmen blasted away at the Jurassic limestone, a living creature emerged from the shattered rock, moved its wings in a desultory fashion, squawked, and fell dead. The men were convinced that it lived for a moment or two after they had blasted the rock inside which it had apparently been encapsulated for millions of years. The *Illustrated London News* for February 9, 1856, reported that a palaeontologist had identified it as a small pterodactyl.

Ambroise Paré, a royal surgeon during the sixteenth century, was watching a powerful workman breaking up large stones in a quarry at Meudon. Both men were astonished when a large toad hopped unconcernedly from a cavity in the centre of a newly split stone.

In 1818, Dr. E.D. Clarke was supervising a deep chalk excavation in the hope of finding fossils there. At a depth of eighty metres his workers uncovered several fossilized sea urchins along with a few newts that looked as if they were only sleeping. Dr. Clarke put three of them on a sheet of paper in the sunlight, and they started to move. Two died within minutes, but the third became so lively that it escaped. Thoughtfully, Dr. Clarke started examining all the newts in the area, but found none that looked like the specimens that had been discovered deep in the chalk excavation. Shortly afterwards, he was lecturing at Caius College, Cambridge, where he showed the newts to the Reverend Dick Cobbold, a respected naturalist. Cobbold examined them carefully and then said that he had never seen any newts like them before. In his opinion they were a primitive species, now extinct.

During the construction of a cellar in the 1860s in Spittlegate, Stamford, United Kingdom, a large, lively toad emerged from solid bedrock more than two metres down. A detailed report of its discovery appeared in the *Stamford Mercury* on October 31, 1862. Three years later a similar toad was discovered alive and well inside a block of limestone during work on the foundations of Hartlepool Waterworks. An account of that episode appeared in the *Leeds Mercury* on April 8, 1865. The Reverend Bob Taylor, a local expert in geology, gave his opinion that the limestone capsule containing the toad was several million years old.

There are several reports of toads emerging from rock strata and coal seams that were laid down millions of years ago. How did they get there? How did they survive?

Victor Loret (1859–1946) was director of the Egyptian Antiquities Services when he carried out some excavations at Saqqara towards the end of the nineteenth century.

Among other interesting finds, the dig turned up a model bird that was more like a working model of a glider than a simple copy of a real bird. The model from Saqqara was sent to the Cairo Museum, where it became catalogue item number 6347. In the late 1960s, however, it was examined in detail by Dr. Dawoud Khalil Messiha, a specialist in radiesthesia (the ability to detect radiations with the human body) and various alternative therapies. He found that the wings were straight and flat like airplane wings, while the tail was vertical, unlike the horizontal tails of normal birds. Furthermore, the Saqqara model had no feet! Dr. Messiha made a replica in balsa wood and found that it glided effectively. The question remains: was it just an ingenious toy glider, made to look like a bird — or was it a model of a large glider that had once carried ancient Egyptian passengers?

Some remarkable golden artifacts from Venezuela, Peru, Costa Rica, and Colombia could also be model aeroplanes, yet they were made by the Mochica or Chimú people whose culture dates back for a good two thousand years. One in particular drew the attention of Ivan Sanderson

Is this Chimú jewellery simply a copy of an insect with wings? Or had the Chimú and Mochica people seen aircraft two thousand years ago?

(1911–1973), who was the leader of the Society for the Investigation of the Unexplained in the United States when the Colombian government arranged for a peripatetic exhibition of their Chimú craft work. It could, arguably, be merely an early representation of a flying insect, but its resemblance to a modern delta-winged aircraft is too close to dismiss it without very thorough investigation. It presents a similar challenge to that offered by the Egyptian bird-glider.

Another mysterious, if rather controversial, discovery was made in February 1961 by three friends: Mike Mikesell, Wallace Lane, and Virginia Maxey. Part of their business was to sell interesting and unusual mineral samples in their retail outlet in Olancha, California, and they had climbed a nearby peak looking for whatever they could find that might be saleable. Sawing through what they thought was a geode that they had picked up near the summit, they encountered a very hard white substance inside it.

The supposed geode was unusual. Inside the porcelain-like substance was a small circle of bright metal; and further examination and X-rays showed more metal at either end. A geologist who examined it at the request of the finders thought that it would have taken hundreds of thousands of years to form. While trying to assess what it might be, or where it might have originated, the finders thought it could even have been made in Atlantis. An alternative hypothesis suggested that it was a spark plug from an internal combustion automobile engine, and a development of this theory said that part of the Olancha "geode" actually bore a close resemblance to a type of spark plug manufactured in the 1920s.

What are the explanations for it? Was it something a time traveller had lost half a million years ago in Olancha? Was it part of an advanced communicator? Was it even one small component of the mechanism of a sophisticated time machine? The same intriguing considerations apply to it as to other anachronistic artifacts: if it is a genuine anachronism, it could be something left behind by a time traveller, an inter-dimensional voyager, or a visitor from another probability track. It could also be evidence that an advanced technology from Atlantis coexisted with pre-technological human hunter-gatherers.

Yet more turbulent controversy seethes around the work of Dr. Javier Darquea (1924–2001) and his Ica stone museum, close to two hundred miles northwest of Lima in Peru. Known as the Museo de Piedras Grabadas (the Museum of Engraved Stones) it contains approximately twenty thousand examples, which the doctor collected between 1966 and 2001. The first specimen was a birthday present given to Dr. Darquea by a farmer of his acquaintance. The doctor was impressed because the carving on the stone looked to him like an extinct species of fish. According to Darquea's testimony, he interrogated a number of knowledgeable, older people in Ica, who gave him clues that pointed him in the direction of a secret cave where he reported that there were over one hundred thousand engraved stones concealed. This cave was said to be located among the Peruvian coastal mountains.

Controversy and suspicion were intensified by Darquea's refusal to lead a team of archaeologists to his "secret cave." Examining mysterious old objects in their original situation can be revealing for experienced archaeologists.

There was some evidence that the farmer who first introduced the mysterious carved stones to Dr. Darquea was accused of selling ancient Peruvian treasures. He promptly confessed to *making* them — in order to avoid the criminal charges associated with selling parts of the Peruvian heritage. Cheating tourists by making and selling forgeries was regarded as only a minor offence.

An interesting historical argument for the genuineness and age of the Ica stones comes from the visit to Peru in 1525 of a Jesuit missionary scholar named Padre Simón. He was one of the party when Pizarro visited Peru. While there, Simón found some engraved stones like the ones Dr. Darquea would collect three hundred years later. Padre Simón's stones were sent to Spain in 1562 — almost forty years after he found them. If the Ica stones are genuine anachronistic artifacts, they are truly remarkable ones. Dr. Darquea saw modern scientific instruments like telescopes cut into them. He also saw various surgical techniques illustrated on the stones.

The Acambaro artifacts present almost the same degree of speculation. In the mid-1940s Waldemar Julsrud, who had moved into Mexico from Germany, claimed to have discovered a number of curious clay models that could have been meant to represent dinosaurs. Radio carbon dating of some organic remains on the surfaces of these models allegedly

Admiral Piri Re'is had a very strange old map. Was it drawn by a time traveller?

showed a date of around 5000 B.C. Charles Hapgood (1904–1982), a pioneer researcher into the mysteries associated with the Piri Re'is map, felt that the Acambaro artifacts of Mexico were genuinely old and mysterious. In Hapgood's opinion they had been intended to represent dinosaurs or unknown ancient creatures long extinct. Would that imply that those who had made the models had actually *seen* the creatures? How did Mexican artists, sculptors, and model makers from 5000 B.C. learn about creatures that predated them by millions of years? It is not beyond the bounds of possibility that along some strangely convoluted spasm-track in time, the sculptors and dinosaurs saw one another.

The mysterious Ica stones that so intrigued Dr. Darquea are matched by the mystery of the curious reliefs carved into the walls of the Temple of Hathor at Dendera in Egypt. Physicists and other electrical specialists tend to interpret these enigmatic carvings as electron tubes with symbolic wave representations inside them. Conventional Egyptologists, however, accept them as characteristic symbols of the sun-boat in which the god Ra made his cosmic journey.

Other interesting examples of what *might* be anachronistic artifacts are the so-called Lion Pillars of India. The one which usually arouses researchers' attention stands near Delhi and seems to have been erected by King Chandragupta II (376–415 A.D.). It's in the region of twenty-three feet high with a diameter of eighteen inches (7 m high and 0.4 m diameter). It weighs around six tonnes. Some researchers have estimated its age at four thousand years, but the consensus makes it much younger at a relatively youthful sixteen hundred years. An earlier ruler, Asoka Vardhana (272–230 B.C.) also raised polished iron pillars in various places. The mystery associated with them is their relative resistance to rust and chemical corrosion. Part of their unusually good state of preservation may be the result of the purity of the iron from which they are made. It also needs to be remembered that the warm, dry air, characteristic of the location, would tend to protect the iron from rust and corrosion. But while making these allowances for the non-mysterious nature of the corrosion-free iron pillars, it must also be borne in mind that they are *abnormally* well preserved, and perhaps some mysterious ancient knowledge from Lemuria or Atlantis reached the makers of the enigmatic Asoka pillars.

If the iron pillars' lack of rust and corrosion can be accounted for without necessarily having recourse to the enigmatic time theories associated with anachronistic artifacts, the crystal lens from Nineveh is not so easy to explain. Austen Henry Layard (1817–1894) was carrying out an archaeological dig at ancient Nineveh when he found a double convex crystal lens in a layer dating from around 600 B.C. Traditional archaeologists prefer to regard Layard's remarkable find as a piece of jewellery or an ornament of some description. It cannot be denied that the crystal lens might merely have been an attractive ornament, but the likelihood of its having been designed and used as a magnifying device cannot be ignored.

In 1829, workers cutting marble eighteen metres down at the Henderson Quarry, not far from Norristown near Philadelphia, Pennsylvania, United States, found a block of marble dating back 8 or 9 million years. Marble is a metamorphosed (changed) rock that started its sedimentary career as limestone and was converted by heat and pressure into denser, harder marble. Some of the workmen sawed through the block that they had noticed and found two Greek letters, π and ι, inscribed inside it. It rather looked as if someone had engraved those letters on the Norriston marble long before homo sapiens had supposedly appeared on the Earth and developed written language. Quarries seem to be useful sources of anachronistic phenomena; the one in Raxton produced an artifact even more puzzling than the one in Norriston.

In June 1844, quarrymen working at Rutherford Mill in Raxton, in southern Scotland, found a length of gold thread embedded in carboniferous rock 2.5 metres down. That particular rock is at least 300 million years old, and the presence of a manufactured gold thread in it raises the familiar questions. How did it get there? Time travel?

Dimensional travel? Probability track travel? Left behind accidentally by extraterrestrial visitors or by inhabitants of Lemuria or Atlantis?

Maximilien Melleville was a distinguished scientist who was working near Laon, the capital of the Aisne district of France, in 1857. The anachronistic artifact that Maximilien discovered was a remarkable chalk globe in a bed of Eocene lignite that was around 50 million years old. The ball did not look as though it was a natural geological phenomenon. Someone, or some*thing*, appeared to have made it deliberately. Why was it lying in an Eocene layer vastly older than the Earth's first human inhabitants?

As a peripheral adjunct to the chalk sphere mystery, Laon is famous for its Chapelle des Templiers, a magnificent twelfth-century octagonal Templar chapel. The Cathedral de Notre-Dame in Laon dates from the same Templar period and seems to have been the model for the sublime architectural work in Chartres, Reims, and Notre-Dame in Paris. The Templars knew many strange secrets, and their presence in the vicinity of the mysterious chalk sphere of Laon may be more than coincidence.

In 1885, an Austrian worker named Reidl broke open a large block of coal that had been brought over from Wolfsegg to the Braun foundry in Schöndorf where Reidl was employed. Inside the coal he found a curious metal cuboid. It was clearly an artifact, and seemed to consist of cast iron. It weighed close to a kilogram and was roughly 6 centimetres by 6 centimetres by 4 centimetres, giving it an approximate volume of 150 centimetres. Close analysis of it in the 1960s revealed that it lacked the nickel, cobalt, and chromium that would have suggested meteoric origin. Like the chalk sphere of Laon, the iron cuboid of Austria has no easy explanation. What was a cast-iron artifact doing inside a piece of coal that dated back around 60 million years?

A similar anachronistic artifact mystery may be linked to the curious appearance of metallic globes or spheres in the pyrophillite mines, sometimes called the wonderstone mines, in Ottosdal in South Africa. (Wonderstone is the popular name for pyrophillite.) The content of these mines is very old indeed, going back to pre-Cambrian times more than 2 billion years ago. Two kinds of the strange metallic spheres have been reportedly found in the mines on numerous occasions. Type one is described as a bluish metal; type two has a spherical metal coating enclosing an absorbent, spongy material. The pyrophillite inside which the metallic spheres occur is a form of soft hydrous aluminium silicate with the chemical formula $Al_2O_34SiO_2H_2O$. The metallic casings of the spheres with the spongy interiors have been analyzed by geologists as comprising pyrites and goethite. Iron pyrites is iron sulphide with the formula FeS_2. The most interesting of its properties is that it exhibits negative resistance when electric current is passed through it. This means that over some ranges of voltage, the greater the voltage, the smaller the current — the exact opposite of what normally happens. It is possible, therefore, that the negative resistance of the pyrites could have something to do with the formation of the anomalous, metal-coated spheres found in the pyrophillite mines at Ottosdal. The goethite content of the metal spheres is equally interesting. Goethite is iron oxy-hydroxide with the formula $FeO(OH)$ and its colour varies through reddish to yellowish dark brown. As a pigment it was known and used by prehistoric cave painters — including those who worked in the Lascaux Caves in France. So how did the mysterious metallic spheres form in the pyrophillite? Did someone, or something, put them there millions of years ago? Have they formed naturally, or are they anachronistic artifacts? Could their appearance be linked to the physical and chemical properties of the pyrites and goethite of which they are made?

The Kabwe Skull — estimated to be about two hundred thousand years old — is sometimes referred to as the Broken Hill Skull. It may be part of what may have been an ancestor of homo sapiens called homo rhodesiensis. It was found in the early 1920s by Tom Zwiglaar, a Swiss miner working in what was then Rhodesia. He sent it to Sir Arthur Smith Woodward (1864–1944) a distinguished paleontologist and one-time president of the Geological Society. Interest in the skull centres on what some researchers have claimed is a bullet hole in it, suggesting that someone in Africa had firearms two hundred thousand years ago. There is certainly a small round hole in the Kabwe Skull, but there are arguments about whether it exited at the other side, or even whether it had begun to heal around its edges. The question as to whether it was a trauma of some kind — such as a bullet, or even a very fast sling stone — that caused the death of homo rhodesiensis remains open; but if pathological skill ever reaches a level where it can answer with objective, scientific certainty that it *was* a bullet, then the Kabwe Skull will be the pivot of one of the most challenging unsolved mysteries that the authors have yet encountered.

If the Kabwe Skull is a source of controversy and conjecture, the Mesoamerican crystal skulls are just as interesting. The best known of these crystal skulls was said to have been found in Lubaantum in Belize by Mike Mitchell-Hedges (1882–1959). Some claims date the skull at close on four thousand years — others believe it belonged to the ceremonial period at Lubaantum, which ended just over one thousand years ago. Frank Dorland, an expert crystal carver himself, examined the Lubaantum skull in the 1970s, and his tests seemed to suggest that it had been cut into its rough shape first and then finished laboriously by grinding it with sand and water. All kinds of extravagant paranormal claims are made about the powers of the crystal skulls. Some researchers maintain that they have heard the skulls growl and chant. Others say that they have miraculous healing powers. Some claim that they have seen pictures inside the skulls — like illuminated thoughts. In other witnesses, some skulls have produced stark, irrational terror, so that they dare not stay in a particular skull's vicinity. Those who support the traditions attached to the skulls and their history argue that they came from outer space at the same as Atlantis was flourishing and were stored in Atlantis before being taken to Mesoamerica. Clearing away some of the wilder speculations and the more extravagant claims still leaves plenty of room for genuine mystery in connection with the crystal skulls. They are of very fine workmanship, and it seems difficult to accept that the techniques for creating them would have been available to early artisans. They may

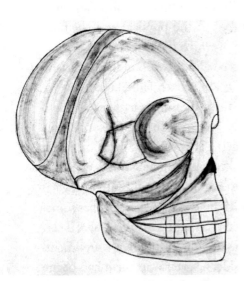

What's the truth behind the mysterious crystal skulls, one of which was reportedly found by Mike Mitchell-Hedges in Lubaantum in Belize? Are they connected with time travel?

indeed have come from *elsewhere.* They might have been created in the nineteenth or early twentieth centuries and "planted" in Mesoamerica as a lucrative hoax. There are grounds for suspicion about their origins: there are also grounds for regarding them as genuine anachronistic artifacts that present a unique challenge to serious, scientific researchers.

The examples of anachronistic artifacts that we have surveyed here *could* be important indicators that help the ongoing investigation into the mysteries and secrets of time. Some may be hoaxes, some may be misunderstandings, some may be misrepresentations … but if only one or two of them prove to be genuinely anachronistic artifacts, they will be of enormous significance.

Chapter 14

THE UNIFIED FIELD THEORY

OF TIME

PHYSICISTS HAVE striven for years to find a unified field theory, one that would bring together all of the fundamental forces and what might best be described as the interactions among the elementary particles. Their hope is to devise one all-encompassing, single theory, including and catering for everything. Einstein worked hard for many years to bring his relativity theories into harmony with theories of electromagnetism — but despite his brilliance, he never quite succeeded.

One major problem is that the forces that act upon objects of normal size are influenced by, or mediated by, fields; however, once the researcher gets down to subatomic dimensions, quantum fields seem to take over and things begin to behave according to the rules of quantum mechanics. Another way of describing what seems to be happening down in those zones occupied by the minutest things is to talk about *exchange particles* that look as if they are able to transfer momentum and energy among things.

One image of what's happening down there is to imagine it as a game of pool played on a curved table with no pockets, on which every ball ricochets off every other ball — exchanging its energy for momentum and vice versa. Extending this pool game metaphor, the unified field theory seeks to explain the relationships of the area in which the pool table sits; the fans in the audience; the shape, structure, and surface texture of the pool table; the players; and the cues and the pool balls in motion. It seeks to link the rules of pool; the interaction of eye, hand, and cue; and why people paid to come to watch the tournament.

In physics, the unified field theory tries to bring four forces into one unified whole. First comes the strong nuclear force, the one that grips quarks tightly together to make protons and neutrons — and also holds those neutrons and protons together to form the nuclei of atoms.

Second is the weak nuclear force, which lies at the back of radioactivity. It seems to repel things at short range and influences quarks, neutrinos, and electrons. Third is the electromagnetic force, which works on charged particles and has the photon as its exchange particle. The fourth force is gravity. It has long-range powers and works on all particles. Its own exchange particle is called the graviton. Because of their very considerable differences, it is no easy matter to unite these four physical forces into one all-inclusive field theory.

In view of time's apparent differences, paradoxes, and anomalies, trying to construct the equivalent of a unified field theory of time is an equally daunting task.

The first obstacle that needs to be surmounted is the problem of the actual *nature* of time. We live in it; we work, play, and relax in it; we eat, sleep, dream, and think in it — yet we have no clear, definitive understanding of what this mysterious phenomenon we call time really *is*. We have devised extremely accurate and complex systems of units and sub-units, and we have technical equipment for calibrating it — but we have no idea of the true *nature* of whatever it is that we are measuring so finely.

The second problem is the conundrum about time's own duration. Dunne put forward some intriguing theories, including infinite regression, and his descriptions of such associated regressions as time one, time two, and time three, up to and including time infinity, cannot be lightly dismissed. An infinite regression in line with Dunne's theories argues powerfully for an infinite time with no beginning and no end. But can it also be argued that everything in our human knowledge and experience exists within time? Dunne's hypothesis supposes that we *observe* time one from time two, time two from time three, and so on. If he's right, then his argument for the infinitude of time is a strong one. But there are equally strong cases to be argued for time having come into existence at some point corresponding to the creation of the universe — if the universe is indeed a creation with a starting point rather than an inexplicable phenomenon that has always existed and will always exist. To our finite human minds, it seems difficult, if not impossible, for time to be both finite and infinite simultaneously: it would not be impossible within the omniscience and omnipotence of an eternally creating and sustaining God.

Then there is the recurring time hypothesis to consider, as well as the circular time concept. Further problems occur when considering the flexibility of time: can it be bent around or twisted so that — somewhere among its entanglements — the past, present, and future apparently coexist from the observer's point of view? The Neanderthal hunter-gatherer, the Greek philosopher, and the twenty-first-century astronaut all see one another and are bewildered by what they see. It is also necessary to try to accommodate the idea of what is perceived as unidirectional, arrow-line flow of time.

Is time fragile? Can it suffer the equivalent of violent, dramatic movements in its counterpart of the tectonic plates that make up Earth's lithosphere? If such warps and glitches happen periodically in time, what effects do they have on intelligent observers — human or otherwise — who inadvertently become involved in them?

To what extent can it be argued that time may not exist at all — except in the mind of the observer? Is it possible that time is merely a subjective phenomenon? Could that explain the apparent super-longevity of characters like Cartophilus and the Count of St. Germain? Did they just have sufficient mental power to refute time, and so to stay young for centuries? Can that mental power be learned and developed? Do we all have the potential to use it? Can mind control matter? There is some evidence that it can. Might this be the vital clue to the next advance in delaying and reversing the human ageing process?

Then there are the enigmatic time portal theories. Are there some amazing ways by which it is possible to locate such doorways into and out of time and to step through them when we wish? Picture time as the external, moving, circular platform of a carousel, revolving around a solid, fixed interior. The showman who knows how to operate the mechanism can move from the central platform to any part of the revolving disc on which the horses and riders move relative to the fixed interior.

How can we combine these very disparate ideas into a unified field theory of time?

Beginning with our Fanthorpean use of the term *instanton*, as explained in Chapter 1, we can regard that as time's equivalent of the exchange particle of quantum physics. If time is to be understood as part of the "theory of everything," then these instantons are the energy and momentum spheres on its pool table. The apparent subjectivity of time — the observer's mental awareness of it — can then be compared to physical relativity. Two children inside a moving train are throwing a softball from one to the other. From their point of view, the ball is travelling at only 3 kilometres an hour within the reference frame of their compartment. From the point of view of an observer standing on the platform as the train passes through the station at 100 kilometres an hour, the ball is travelling first at 103 kilometres an hour as it is thrown in the direction in which the train is travelling, and then at 97 kilometres an hour as it is thrown in the opposite direction. The children in the train and the observer on the platform are all making true and accurate observations from their different frames of reference. If an observer from space now adds a further frame of reference to the situation, that observer sees the Earth rotating on its axis and revolving around the sun, while the sun and its solar system move through the void. From the astronaut's point of view the softball inside the train is moving along an extremely complicated path at several thousand kilometres an hour. If we can change our frames of reference, we can exchange our experiential realities. If the ball somehow escaped from the train and struck the astronaut's ship at 100,000 kilometres an hour, it would destroy them both. If the astronaut were sharing the compartment with the children, the ball would land lightly in his hand and he would return it to the children with a gentle smile. In a less spectacular — and rather more probable way — if the ball flew through the train window and hit the observer on the platform it might cause bruising and discomfort, but would not have the same consequences as an impact with the spaceship at 100,000 kilometres an hour. As part of a unified theory of time, it may be argued that when time references are changed, the consequences for those experiencing such changes can range from almost imperceptibly mild to traumatically severe.

The subjectivity of time (allied to that frames-of-reference concept) may account — at least to some extent — for the observation of time as having different impact effects for observers in different frames of reference. The one-day life experience of a minute insect may be the equivalent of a human century from that ephemeral insect's frame of reference. That day, from the frame of reference of a bristlecone pine that has already lived four thousand years or more, may be like one single tick of a clock. Observational subjectivity may also account for the apparent differences in the nature, substance, or fabric of time. But objective time — if it exists — may also be thought of as possessing a chameleon-like ability to change its appearance and its observed tempo so that it can stand still, walk, or run for different observers in different situations.

The authors' unified field theory of time seeks to accommodate the reciprocity of both objective and subjective time. Time may *seem* to behave very differently from different observers' varied frames of reference, but — like the subcutaneous body of the colour-changing chameleon — there is a real, objective *something* underneath the changing colours of time. At those different levels of its existence, the chameleon can be thought as having the ability to be both the same and different. Our unified field theory of time suggests that — inseparable from the subjective observations of time — there is a consistent and mysterious objective something.

Is it possible to unify the concept of unidirectional, arrow-line time with circular, recurring time — whether time is finite and terminable or infinite and interminable? Imagine a circle around a sphere, like the equator running around the Earth. Provided that the observer is capable (because of human mortality) of traversing only a small fraction of that circular path, it will seem indistinguishable from a straight line as far as that observer is concerned. However, should phenomena such as reincarnation and déjà vu actually exist, then a long enough succession of lives — even with only an impartial memory of each — may suggest to the observer that there is something akin to circularity or cyclic repetition in his experience of time

The unified field theory of time also needs to take into account the theory of bifurcated time and probability tracks: experiential versus non-experiential Worlds of If. When an opportunity arises to make a decision, the observer's ensuing experiential life track will be the one that is consequential upon the choice that was made.

How can these bifurcated time theories fit in with those other aspects of time that are germane to our unified field theory? Probability track bifurcations can separate out from circular time as easily as they can diverge from unidirectional time. Each bifurcation may, in fact, eventually form its own probability circle, running alongside the original one. Supposing, as well may happen, that these tangential Worlds of If recombine with the original time cycle at some later point, the observer's sense of déjà vu — together with reincarnation evidence — will be reinforced.

If — as cases such as the Versailles adventure of Moberly and Jourdain may indicate — time is convoluted and can bend back on itself, how does that square with our unified field theory of time? The evidence from anachronistic artifacts also suggests that time may occasionally double back on itself, creating riddles for us in the ancient rock strata. Whether it's cyclic or linear, the substance from which time is formed may well prove to be capable of bending back on itself, warping, twisting, and tangling until what is normally regarded as past, present, and potential future probabilities are brought closely enough together to be mutually visible to observers in all three time zones.

Evidence for déjà vu and reincarnation is persistent enough not to be discarded, and might be said to point in the direction of cyclic time. Daily, common-sense experiences and pragmatic observations pointing to unidirectional arrow-flow time are too strong to ignore. Startless and endless eternal time is a phenomenon that human observers are not yet in a position to measure, but J.W. Dunne's infinite regression theories are well worth considering.

How can it all be put together into a rational and coherent whole? Our unified field theory of time begins at the microcosmic level of what we have termed instantons — the basic exchange particles of time — where interchangeable teleological energy and momentum dance exuberantly together. The substance of time, which is made up of these cavorting instantons, is infinitely plastic and flexible. Although we, as time observers, think that for most of our sequential existences we are moving in a unilinear direction, the locus along which we believe that we are experiencing forward movement is capable of such major convolutions that we can occasionally observe scenes from the actual past and from one or more future probability fields. This almost infinitely long, convoluted locus, with its wealth of bifurcating probability loci, may well be cyclic or circular — and may equally possess portals through which observers can enter and leave, producing the St. Germain and Cartophilus phenomenon. Finally, the subjective qualities of time also need to be included in this unified field theory: so we return to the metaphor of the chameleon. The observer may well have sufficient mind power to exert significant influence over the outward appearance and functioning of time within any given frames of reference — but there is nevertheless an objective inner core of time that may prove much harder to control.

BIBLIOGRAPHY

Andrews, William. *Curiosities of the Church*. London: Methuen & Co., 1890.

Andrews, William. *Antiquities and Curiosities of the Church*. London: William Andrews & Co., 1897.

Asimov, Isaac. *Guide to Earth and Space*. New York: Fawcett Crest, 1991.

Asimov, Isaac. *The Universe*. New York: Avon Books, 1971.

Baring-Gould, S. *Historical Oddities and Strange Events*. London: Methuen and Co., 1890.

Boslough, John. *Stephen Hawking's Universe*. Great Britain: Fontana Paperbacks, 1986.

Burstein, Dan, ed. *Secrets of the Code*. London: Orion Books, 2005.

Bernstein, Jeremy. *Einstein*. Glasgow: Fontana, 1978.

Brennan, J.H. *Time Travel*. U.S.A.: Llewellyn Publications, 2003.

Brookesmith, P., ed. *The Enigma of Time*. London: Orbis Publishing, 1984.

Cavendish, Richard, ed. *Encyclopaedia of the Unexplained*. London: Routledge & Kegan Paul, 1974.

Clarke, Arthur C. *Mysterious World*. London: William Collins Sons & Company, 1980.

Clark, Jerome. *Unexplained*. USA: Gale Research Inc., 1993.

Clark, Robert E.D. *The Universe; Plan or Accident*. London: The Paternoster Press, 1949.

Crane, Peter. *Miracles and Modern Science*. Great Britain: Bourne Press Ltd., 1991.

Darling, David. *Equations of Eternity*. New York: Hyperion Books, 1993.

Davies, Rodney. *Doubles*. London: Robert Hales Ltd., 1998.

Dennis, Felix. *How to Get Rich*. London: Ebury Press, 2006.

Dunford, Barry. *The Holy Land of Scotland*. Scotland: Brigadoon Books, 1996.

Dunne, J.W. *An Experiment with Time*. London: Faber and Faber Limited, 1927.

Dyall, Valentine. *Unsolved Mysteries*. London: Hutchinson, 1954.

Eysenck, H.J & Carl Sargent. *Explaining the Unexplained*. London: BCA, 1993.

Encyclopaedia Britannica: *Britannica Online*: http://www.eb.com.

Fanthorpe, Patricia & Lionel. *The Holy Grail Revealed*. California: Newcastle Publishing Co. Inc., 1982.

Fanthorpe, Lionel & Patricia. *The Oak Island Mystery*. Toronto: Hounslow Press, 1995.

Fanthorpe, Lionel & Patricia. *Secrets of Rennes le Château*. U.S.A.: Samuel Weiser Inc., 1992.

Fanthorpe, Lionel & Patricia. *Mysteries of Templar Treasure & the Holy Grail*. U.S.A.: Red Wheel/Weiser, LLC., 2004.

Fanthorpe, Lionel & Patricia. *The World's Greatest Unsolved Mysteries*. Toronto: Hounslow Press, 1997.

Fanthorpe, Lionel & Patricia. *The World's Most Mysterious People*. Toronto: Hounslow Press, 1998.

Fanthorpe, Lionel & Patricia. *The World's Most Mysterious Places.* Toronto: Hounslow Press, 1999.

Fanthorpe, Lionel & Patricia. *Mysteries of the Bible.* Toronto: Hounslow Press, 1999.

Fanthorpe, Lionel & Patricia. *The World's Most Mysterious Objects.* Toronto: Hounslow Press, 2002.

Fanthorpe, Lionel & Patricia. *The World's Most Mysterious Murders.* Toronto: Hounslow Press, 2003.

Fanthorpe, Lionel & Patricia. *Unsolved Mysteries of the Sea.* Toronto: Hounslow Press,. 2004.

Fanthorpe, Lionel & Patricia. *Mysteries and Secrets of the Templars.* Toronto: Hounslow Press, 2005.

Fanthorpe, Lionel & Patricia. *The World's Most Mysterious Castles.* Toronto: Hounslow Press, 2005.

Fanthorpe, Lionel & Patricia. *Mysteries and Secrets of the Masons.* Toronto: Hounslow Press, 2006.

Foreman, Joan. *The Mask of Time.* London: Macdonald and Jane's, 1978.

Gamow, George. *One Two Three ... Infinity.* London. Macmillan & Co Ltd., 1962.

Grant, Joan. *Winged Pharaoh.* U.S.A.: Alliance Press, 1989.

Gribbin, John. *White Holes.* Herts., U.K.: Paladin, 1977.

Guirdham, Arthur. *The Lake & The Castle.* Cygnus Books, 1992.

Guirdham, Arthur. *We Are One Another.* Cygnus Books, 1992.

Guirdham, Arthur. *The Cathars and Reincarnation.* London: Neville Spearman. 1970.

Hill, G. *Secrets of the Unknown.* Great Britain: Hodder and Stoughton Ltd., 1979.

Hurkos, Peter. *Psychic.* London: Arthur Barker Limited. 1962.

Lewis, C.S. *The Abolition of Man.* Great Britain: Fount Paperbacks, 1978.

Lomas, Robert. *The Invisible College.* London: Headline Book Publishing. 2002.

Jastrow, Joseph. *Error and Eccentricity in Human Belief.* New York: Dover Publications Inc., 1935.

Krassa, P. *Father Ernetti's Chronovisor.* Florida: New Paradigm Books, 2000.

Lemesurier, Peter. *Beyond all Belief.* Great Britain: Element Books Ltd., 1983.

Lowestoft College Magazine, Number 11, Midsummer Term 1910, printed by Flood and Son Ltd, the Borough Press, Lowestoft.

Mack, Lorrie et al, ed. *The Unexplained.* London: Orbis, 1984.

Marten, M. and J. Chesterman. *The Radiant Universe.* Dorset: Blandford Press, 1980.

Massingham, Hugh & Pauline. *The London Anthology.* London: Spring Books, 1959.

Matthew, Caitlin. *Sophia: Goddess of Wisdom.* Quest Books, Theosophical Publishing House, 1991.

McEvoy, J.P. and O. Zarate. *Stephen Hawking for Beginners.* Cambridge: Icon Books, 1995.

Miller, G.H. *Dictionary of Dreams.* Bath: Parragon Books, 1999.

Morison, E. and Lamont, F. [Moberly, Charlotte Anne & Jourdain, Eleanor]. *An Adventure.* London: Macmillan and Co Ltd., 1913.

Nicholas, M. *The World's Greatest Psychics and Mystics.* London: Hamlyn Publishing Group, 1994.

Penrose, Roger. *The Road to Reality.* London: Vintage Books, 2005.

Playfair, G.L. *The Unknown Power.* Granada Publishing Ltd., 1977.

Priestley, J.B. *Man and Time.* London: Aldus Books Limited, 1964.

Reader's Digest Book. *Folklore, Myths and Legends of Britain.* London: The Reader's Digest Association Ltd., 1973.

Reader's Digest Book. *Strange Stories, Amazing Facts.* London: The Reader's Digest Association Ltd., 1975.

Rolleston, T.W. *Celtic Myths and Legends.* London: Studio Editions Ltd., 1994.

Russell, Eric Frank. *Great World Mysteries.* London: Mayflower, 1967.

Saltzman, Pauline. *The Strange and the Supernormal.* New York: Paperback Library Inc., 1968.

Sampson,Chas. *Ghosts of the Broads.* Norwich: Jarrold Colour Publications, 1973.

Sharper Knowlson, T. *The Origins of Popular Superstitions and Customs.* London: Studio Editions Ltd., 1995.

Simmons, I. & Quin, M. *Fortean Studies Volume 7.* London: John Brown Publishing Ltd., 2001.

Singh, Simon. *The Cracking Codebook.* London: HarperCollins Publishers Ltd., 2001.

Spencer, John & Anne. *The Encyclopaedia of the World's Greatest Unsolved Mysteries.* London: Headline Book Publishing, 1995.

Wilson, Colin. *The Psychic Detectives.* London: Pan Books, 1984.

Wilson, Colin. & Wilson, Damon, *Unsolved Mysteries Past and Present.* London: Headline Book Publishing Plc., 1993.

Wilson, Ian. *The After Death Experience.* London: Corgi, 1987.

Wilson, Ian. *Reincarnation.* Harmondsworth, England: Penguin Books Ltd., 1981.

Young, George. *Ancient Peoples and Modern Ghosts.* Nova Scotia: Young, 1991.

Lightning Source UK Ltd.
Milton Keynes UK
UKOW01f0255190915

258862UK00008B/113/P